Evolutionary Biology

VOLUME 33

A Continuation Order Plan is available for this series. A continuation order will bring delivery of each new volume immediately upon publication. Volumes are billed only upon actual shipment. For further information please contact the publisher.

Evolutionary Biology

VOLUME 33

Edited by

ROSS J. MACINTYRE

Cornell University
Ithaca, New York

and

MICHAEL T. CLEGG

University of California, Riverside
Riverside, California

SPRINGER SCIENCE+BUSINESS MEDIA, LLC

The Library of Congress catalogued the first volume of this title as follows:

Evolutionary biology, v. 1– 1967–

New York, Appleton-Century-Crofts.
v. illus., 24 cm annual.
Editors: 1967– T. Dobzhansky and others.
1. Evolution—Period. 2. Biology—Period. I. Dobzhanksy, Theodosius Grigorievich, 1900–

QH366.A1E9 575′.005 67-11961

ISSN 0071-3260

ISBN 978-1-4419-3385-0 ISBN 978-1-4757-5190-1 (eBook)
DOI 10.1007/978-1-4757-5190-1

Originally published by Kluwer Academic / Plenum Publishers in 2003

http://www.wkap.com

10 9 8 7 6 5 4 3 2 1

A C.I.P. record for this book is available from the Library of Congress

Contributors

J. Stuart F. Barker • *University of New England, Armidale, New South Wales, Australia*

John A. Birdsell • *University of Arizona, Tucson, Arizona 85721*

D. J. Hosken • *University of Zurich-Irchel, Zurich, Switzerland CH 8057*

Bruce S. Lieberman • *The University of Kansas, Lawrence, Kansas 66045*

Shane F. McEvey • *Australian Museum, Sydney, New South Wales, Australia*

Scott Pitnick • *Syracuse University, Syracuse, New York 13244*

Michal Polak • *University of Cincinnati, Cincinnati, Ohio 45221*

William T. Starmer • *Syracuse University, Syracuse, New York 13244*

P. Stockley • *University of Liverpool, Leahurst, Neston, South Werral CH64 4BP*

Christopher Wills • *University of California San Diego, La Jolla, California 92093-0116*

Larry L. Wolf • *Syracuse University, Syracuse, New York 13244*

Preface

This volume is the thirty-third in this series, which includes 32 numbered volumes and an unnumbered supplement. Several special volumes have also been published as separate monographs. This volume, like the others in the series, has chapters devoted to a broad spectrum of topics. Indeed, the editors continue to solicit manuscripts on subjects covered by the encompassing rubric of Evolutionary Biology. Manuscripts should be sent to either Ross J. MacIntyre, Department of Molecular Biology and Genetics, Cornell University, Ithaca, NY 14853 or Michael T. Clegg, Department of Botany and Plant Sciences, University of California, Riverside, CA 92521.

Table of Contents

Unifying Theory and Methodology in Biogeography

Bruce S. Lieberman

The Evolutionary Origin and Maintenance of Sexual Recombination: A Review of Contemporary Models

John A. Birdsell and Christopher Wills

Evolutionary Biology

VOLUME 33

1

Unifying Theory and Methodology in Biogeography

BRUCE S. LIEBERMAN

INTRODUCTION

Biogeography is a discipline with a long intellectual heritage (Brown and Lomolino, 1998; Browne, 1983; Mayr, 1982; Nelson, 1978) that considers where and why different types of organisms occur over the face of the globe. Almost since its inception, there has been a fundamental debate in the field about how to best explain biogeographic patterns. This debate has centered on two primary explanations. In the dispersalist or mobilist explanation, organisms were thought to have continually moved or dispersed between different regions such that similar types of organisms would be shared between regions due to episodes of dispersal. In the other, the vicariant or extensionist explanation, organisms start out with broad distributions. These distributions would then be subsequently fragmented by the emergence of geological or climatic barriers which concomitantly promote evolutionary divergence in the now isolated populations. In this explanation, it was held that similar types of organisms were shared between regions due to the fact that these regions had a common geological history.

More recently, since the publication of MacArthur and Wilson (1967) and works in cladistic biogeography such as Croizat et al. (1974), a new but related debate has emerged over the relative importance of ecological

BRUCE S. LIEBERMAN • Departments of Geology and Ecology and Evolutionary Biology, University of Kansas, Lindley Hall, Lawrence, KS 66045. blieber@ku.edu

Evolutionary Biology, Volume 33
Edited by MacIntyre and Clegg, Kluwer Academic/Plenum Publishers, 2003

factors versus historical factors as a general explanation of biogeographic patterns (Brooks and McLennan, 1991; Brown and Lomolino, 1998). Ecological biogeography is the subdiscipline of biogeography that focuses on the study of how ecological processes and mechanisms influence the distribution of organisms, whereas historical biogeography is the subdiscipline of biogeography that focuses on how geological events influence the evolution and distribution of groups of organisms. Herein, methodological tools and theoretical principles will be discussed that allow resolution of these debates and unification of these alternative approaches.

HIERARCHIES IN NATURE

The best way to elucidate how historical and ecological biogeography are linked is to contemplate the hierarchical structure of nature. Many evolutionary biologists, particularly paleobiologists, have asserted that nature is made up of entities that are arranged in hierarchies (e.g., Allen and Starr, 1982; Arnold and Fristrup, 1982; Brooks, 1988; Brooks and McLennan, 1991; Brooks and Wiley, 1986; Burns et al., 1991; Buss, 1987; Damuth, 1985; Dawkins, 1982; Dobzhansky, 1937; Eldredge, 1982; Eldredge, 1985; Eldredge, 1989; Eldredge, 1995; Eldredge and Cracraft, 1980; Eldredge and Ormiston, 1979; Eldredge and Salthe, 1984; Feynman, 1965; Goodwin, 1994; Gould, 1980; Gould, 1982; Gould, 1990; Hull, 1988; Hull, 1980; Lieberman, 1995; Lieberman et al., 1993; Lieberman and Vrba, 1995; Mayr, 1942; Mayr, 1982; Rosen, 1991; Salthe, 1985; Smuts, 1925; Sober, 1984; Streidter and Northcutt, 1991; Valentine and May, 1996; Vrba, 1980; Vrba, 1983; Vrba, 1989; Vrba, 1996; Vrba and Eldredge, 1984; Vrba and Gould, 1986; Williams, 1992). The claim that nature is hierarchically structured is based on the fact that there are discrete entities that can be recognized in nature, and further these entities are nested, with smaller entities inside of larger entities. Each level has autonomy, and higher level entities are not merely aggregates of lower level entities (Eldredge, 1985). This autonomy implies properties specific to that hierarchical level, termed emergent properties. For example, a species contains organisms, but it is more than just a simple summation of individual organisms. In sexual species its definition involves some statement about the nature and extent of patterns of interbreeding among different organisms.

In nature there are two major varieties of hierarchically arrayed entities. One major variety is best viewed as genealogical. These include genes, chromosomes, germ-line cells, organisms, reproducing populations (sometimes referred to as demes), species, and clades (see the extensive reference list on hierarchies cited above for equivalent lists). All of these different genealogical entities house some form of information. They also reproduce, making more entities of like kind. They thus develop and modify information, except in the case of higher taxa. These cannot reproduce without becoming paraphyletic and thus no longer clades (Vrba, 1996). These genealogical entities belong to the genealogical hierarchy.

There also exist hierarchically nested entities involved in matter/energy transfer, exchange, and transformation. Such entities include proteins, somatic cells, tissues, organs, organisms, interacting populations (these have been referred to as avatars by Damuth [1985] and Eldredge [1989]), communities and ecosystems, and finally regional biotas and the biosphere. These entities belong to the economic or ecological hierarchy. Within each level, different entities interact with one another.

Analyses of real world data suggest differences between patterns in the economic hierarchy and patterns in the genealogical hierarchy. For example, Mayden's (1988) study of the freshwater fish of North America demonstrated that there are major differences between how communities are structured and organized and the patterns of phylogenetic relatedness among the taxa whose populations built up these communities (see also Brooks and McLennan [1991]). Vrba (1987) recognized another pattern of disjunction among entities of the economic and genealogical hierarchy. Among African antelopes she found that there are two types of clades. One type that is very diverse and another type that is species poor. Within the African ecosystems the species occur in, they show precisely the opposite pattern of abundance. The species in the diverse clade have relatively few individuals within any given ecosystem, whereas the species from the depauperate clade have many individuals in ecosystems.

Because there exists a hierarchically nested set of such genealogical and economic entities, the patterns that we see in these entities when we study them through time may be expected to be broadly hierarchical (Fig. 1). Most of the entities listed belong to only one of the two hierarchies. Within hierarchies there can be upward and/or downward pointing arrows or vectors of causation. The entities in the economic hierarchy are of greatest interest to ecologists, whereas entities in the genealogical hierarchy are of greatest interest to evolutionary biologists. From the discussion above and also Fig. 1, it is clear that there are a few entities, like organisms and populations, that are common to both the genealogical and economic hierarchies. These are the entities which allow vectors of causation to extend across the hierarchies (Vrba and Eldredge, 1984; Eldredge and Salthe, 1984; Eldredge, 1985, 1989; Vrba, 1987).

One of the important conceptual developments engendered by the hierarchical expansion is that it is invalid to extrapolate evolutionary or

CLADES		*BIOSPHERE/ REGIONAL BIOTAS*
SPECIES		*ECOSYSTEMS/ COMMUNITIES*
POPULATIONS (DEMES)	⟷	*POPULATIONS (AVATARS)*
ORGANISMS	⟷	*ORGANISMS*
GERM-LINE CELLS		*ORGANS*
CHROMOSOMES		*SOMATIC CELLS*
GENES		*PROTEINS*

FIG. 1. A list of hierarchically arrayed entities in both the genealogical (left-hand side) and economic (right-hand side) hierarchies. Arrows represent entities found in both hierarchies.

ecological processes operating at a lower hierarchical level to explain patterns in a higher level genealogical or economic entity. Specifically, some theories about the evolutionary process are based on the premise that we can extrapolate processes at one hierarchical level, let's say the genetics of populations, to explain the diversification of a clade. The problems with this extrapolationist approach in evolutionary biology have been discussed extensively by Gould (1980, 1982, 1990). If each of the levels of the genealogical hierarchy represent distinct, real entities, with emergent processes and distinct processes, then processes that operate at one level may not freely translate to another level.

HIERARCHY THEORY AND BIOGEOGRAPHY

That the hierarchical formulation also critically impacts the field of biogeography was first recognized by Brooks (1988) and Brooks and McLennan (1991). If there exist hierarchically structured biological entities, then it is clear that this must relate to biogeography (Brooks, 1988; Brooks and McLennan, 1991). Within both the genealogical and the economic hierarchies there are entities that can display biogeographic patterns. Biogeographic patterns within the genealogical hierarchy will involve entities that reproduce and preserve information, and they will be governed by evolutionary processes. These can best be visualized as historical biogeographic patterns. Patterns within the economic hierarchy will involve entities that undergo matter/energy exchange and transfer. These can best be visualized as ecological biogeographic patterns, though it has not been defined this way traditionally.

It is the existence of the two hierarchies that should set the boundaries between the two subdisciplines of biogeography, historical biogeography and ecological biogeography. However, as traditionally defined ecological biogeography (sensu Brown and Lomolino [1998]) focused on the role of ecological factors in biogeography whereas historical biogeography (sensu Nelson and Platnick [1981]) focused on the role of geological and climatic factors in biogeography. These hierarchies will specify where there should be interdigitation between these two subdisciplines, because some entities are shared between the two hierarchies, specifically, organisms, and potentially populations (to the extent that avatars and demes overlap). This provides one opportunity to unite these different subdisciplines and approaches.

Biogeographic Patterns in the Genealogical Hierarchy

The fundamental pattern in historical biogeography that biogeographers seek to uncover is taxa that show similar patterns of evolutionary differentiation across geographic space. That is, in several different groups, the taxon in one area is always the closest relative of the taxon in another area (Fig. 2). Of course, in the real world things are rarely this simple, but still, the fundamental signal historical biogeographers are seeking to uncover in their data is this pattern. Historical biogeographic patterns could involve the relationship between evolutionary change and geographic distribution for entities at several levels of the genealogical hierarchy including organisms, populations, species, and clades. Because each of these levels has emergent properties unique to them, and also evolutionary processes intrinsic to them, patterns at any one lower level cannot be smoothly extrapolated to any higher level. Patterns at each of these levels should operate at an approximate time scale that is also hierarchical, with the time interval we associate with biogeographic differentiation of higher taxa longer than the time scale we associate with biogeographic differentiation of populations. Further, the evolutionary, geological, and/or climatic processes that would influence biogeographic patterns within species might differ from those that influence patterns among species within clades. For example, some of the

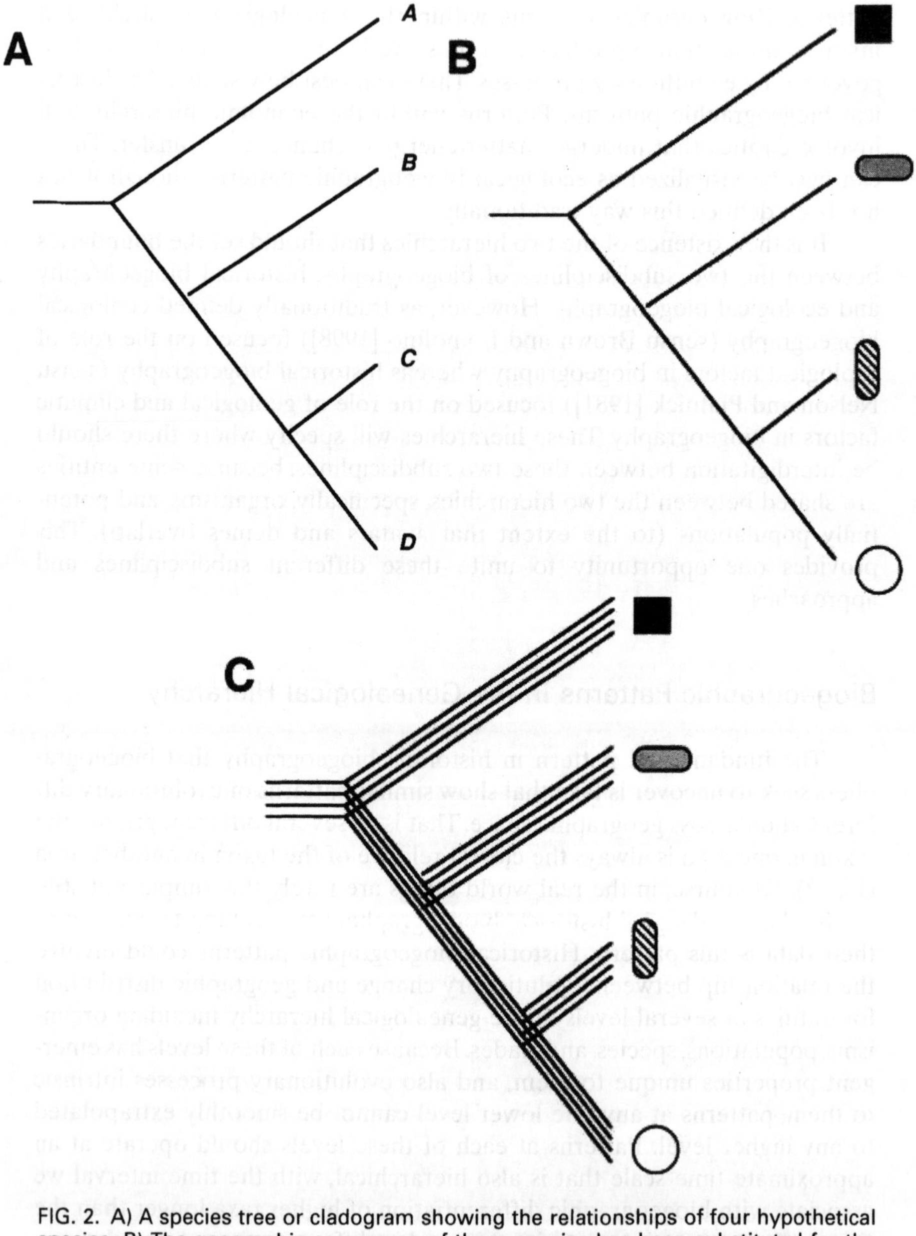

FIG. 2. A) A species tree or cladogram showing the relationships of four hypothetical species. B) The geographic occurrence of these species has been substituted for the taxon name. The different shapes represent different geographic regions. C) Four separate clades showing the same pattern of biogeographic relationship among areas, evidence for congruent patterns of evolution across geographic space.

patterns within clades could be explained by such processes as allopatric speciation, vicariance, and dispersal. Some of these processes might not be immediately applicable to other hierarchical levels because at different spatial and temporal scales different evolutionary and ecological processes may operate (Brooks and Wiley, 1986; Ricklefs, 1987; Brooks, 1988; Brown and Maurer, 1989; Brooks and McLennan, 1991).

Biogeographic Patterns in the Economic Hierarchy

In addition to historical biogeographic patterns in each of the entities in the genealogical hierarchy, there are potentially ecological biogeographic patterns in each of the entities in the economic hierarchy. As an example of these types of patterns, Brown and Maurer (1989) considered how physical space and nutritional resources were divided in the diverse extant North American avian and mammalian clades. Ecological biogeographers can also study changes in the distribution of a single species over short temporal and spatial scales.

Relating Biogeographic Patterns in the Two Hierarchies

As discussed above, the relative prevalence of a signal for each of these two types of biogeographic patterns, ecological and historical, has long been debated. This debate arises because within each of the two hierarchies there are hierarchically arrayed entities, each of which can display some form of biogeographic patterns. Patterns, and the processes that produced them, cannot necessarily be extrapolated between levels of a single hierarchy. What about relating biogeographic patterns across the hierarchies? Brooks and McLennan (1991) argued that as the spatial and temporal scale of any biogeographic study increases, the historical biogeographic signal overrides patterns from ecological biogeography because Brooks (1981), Mayden (1988), and Ross (1972, 1986) demonstrated that the geographic distribution of a species' closest relative is a better predictor of its geographic distribution than its ecological characteristics. However, there are cases where an ecological biogeographic signal may be present and strong at broad spatial and temporal scales (Brown and Lomolino, 1998; Brown and Maurer, 1989; Stevens, 1992).

At the highest levels in the two hierarchies, the contrasts between the entities of the different hierarchies are so profound that comparing biogeographic patterns becomes impossible because we need to consider different types of data. For example, the question of why taxonomic diversity correlates with latitude throughout the biosphere is an ecological biogeographic one. This is a topic that includes genealogical entities, as there are also similarly shaped latitudinal diversity gradients within individual higher taxa too, but it is primarily a topic relating to ecology and entities in the economic hierarchy (Stevens, 1992). Specifically, why can we pack more populations into a tropical ecosystem than into a boreal one? By contrast, a historical biogeographic study might seek to determine if a set of taxa in different clades show similar patterns of speciation associated with the establishment of several geographic barriers. Within the regions being studied, ecosystems will of course be present, but these entities will not be of direct interest and thus will not be sampled as such. Thus, at this level historical and ecological biogeographic patterns are divergent.

However, when we consider biogeographic patterns in smaller entities, for example, biogeographic differentiation within populations, then there may be close congruence between historical and ecological biogeographic patterns. In such a case, the geographic occurrence of organisms within populations related via a pattern of descent and the geographic occurrence of organisms interacting with one another in populations might be similar.

THE RELATIONSHIP BETWEEN CLIMATIC CHANGE, GEOLOGICAL CHANGE, AND BIOGEOGRAPHIC PATTERNS

Other differences emerge between biogeographic patterns in entities of the genealogical and economic hierarchies. Some of these can be conceptualized by considering how climate change influences the different entities in the two hierarchies. There are patterns of climate change governed by distinct astronomical cycles (Huntley and Webb, 1989). These include daily, seasonal, and yearly cycles which in turn influence changes in the behavior of organisms and can lead to large scale migrations. There are also cycles in the abundance of sunspots which influence climate on decadal time scales, and oceanographic changes that influence climate on similar time scales (McGowan et al., 1998). Each of these has concomitant effects on organismal distribution. Then there are even longer term Milankovitch climate cycles operating between roughly 20,000 and 100,000 year intervals (e.g., Hays et al. [1976], Berger [1980], Imbrie and Imbrie [1980], and Huntley and Webb [1989]).

It is worthwhile to focus first on biogeographic patterns in genealogical entities that are attributable to Milankovitch cycles. These cycles operate on time scales much longer than the existence of any organism and probably most if not all populations. Instead, they would most profoundly influence biogeographic patterns within species and also potentially within the clades that those species make up as well as in higher level economic entities. The patterns of climatic change associated with Milankovitch cycles and their influence on geographic distribution have been particularly well documented for the Pleistocene and Holocene epochs. It has been shown by Vermeij (1978) in the case of marine invertebrates, Coope (1979, 1990) in the case of insects, Davis (1986), Huntley and Webb (1989), and Foster et al. (1990) in the case of forest tree species, Graham (1986, 1992) and Graham et al. (1996) in the case of mammals, and Bennett (1990) in the case of marine microfossils, that between glacial and interglacial periods species can come to have dramatically different distributions, all in an effort to track the environment that they prefer.

Often species change little if at all morphologically throughout these episodes of long term climate change. Instead, they simply track their preferred environment (Huntley and Webb, 1989; Bennett, 1990; Vrba, 1993, 1996). However, in many cases, Milankovitch climate cycles not only induce movements of species, but they can also lead to extinction or speciation. This has been described by Vrba (1980, 1983, 1985, 1993, 1996) as the Turnover Pulse hypothesis. The Turnover Pulse operates when the climate changes and species try to track their preferred environment. This can lead to the isolation of populations of these species, which encourages speciation. Sometimes, if the preferred habitat of a species shrinks too much due to climate change, the species will go extinct. In the Turnover Pulse, patterns of speciation and extinction will be replicated across several clades.

Milankovitch climate cycles also influence entities in the economic hierarchy. Populations of the different species that make up communities and ecosystems tend to migrate individualistically and in different directions in response to the environmental changes driven by Milankovitch cycles (Davis, 1976; Huntley and Webb, 1989; Bennett, 1990; Foster et al., 1990; Graham, 1992; Vrba, 1993, 1996). As the climate shifts, populations of different species will track their preferred environmental parameters, often over very long distances. After substantial climate change, the exact combination of environmental parameters needed to sustain all the important populations of a given ecosystem are not likely to be met. The result of this is that species that were once in contact tens of thousands of years ago are no longer in contact, such that ecological communities and ecosystems are completely reorganized and reassembled by Milankovitch cycles. At this scale, in terms of biogeographic patterns produced by Milankovitch cycles, entities in the genealogical and ecological hierarchies show very different responses. Communities and ecosystems are evanescent in the face of these cycles (Huntley and Webb, 1989; Bennett, 1990; Foster et al., 1990; Graham,

1992; Vrba, 1993, 1996) whereas genealogical entities like species and clades can persist and indeed are sometimes even obdurately stable.

Because of this phenomenon, over Milankovitch type time scales it will generally be harder to study biogeographic patterns in entities like an individual community or ecosystem than in a genealogical entity like a clade. This is because at these time scales communities and ecosystems quickly cease to exist and thus cannot show any prolonged pattern; however, there could be general rules or ecological processes associated with the establishment of communities that related in some way to the geography of the region that the communities occupied.

Although there are many cases when historical and ecological biogeographic patterns diverge, there are also cases when they coincide. For example, several species may have overlapping geographic ranges. Patterns of population level differentiation can be examined within each of these species using molecular methods. It may be found that the pattern of geographic differentiation within each of these species is similar, such that populations from one area are always the closes relatives of populations from another area (see Fig. 1). In a historical biogeographic framework, this would be explained by suggesting that geological or climatic factors played a prominent role in influencing intra-specific differentiation in several species (Avise, 1986). If the populations of many of these species were continually interacting ecologically throughout the evolutionary divergence, the biogeographic pattern could be viewed as a coevolutionary phenomenon (Brooks, 1985; Avise, 1986, 1992; Avise et al., 1987; Kluge, 1988; Brooks and McLennan, 1991; Zink, 1991), with entities in both the economic and genealogical hierarchy differentiating along lines of geographic space. Here, there would be concordant historical and ecological biogeographic differentiation, and this is possible because populations can be entities in both the genealogical and the economic hierarchy.

There are characteristic processes that operate on time scales even longer than those of the Milankovitch cycles which can structure biogeographic patterns in entities of the genealogical and economic hierarchies. One set of processes are earth history events mediated by plate tectonics, such as continental rifting and continental collision. These events operate on time scales of many millions of years, and can profoundly influence biogeographic patterns. For example, continental rifting can act to isolate populations of species, thereby facilitating speciation via allopatric speciation and vicariance (Croizat et al., 1974; Platnick and Nelson, 1978; Rosen, 1978, 1979; Brooks, 1981, 1985, 1988; Nelson and Platnick, 1981; Cracraft, 1982; Wiley and Mayden, 1985; Mayden, 1988; Wiley, 1988a, b; Funk and Brooks, 1990; Brooks and McLennan, 1991; Lieberman and Eldredge, 1996). Such plate tectonic events can induce congruent speciation events and differentiation within clades that occur in the same region. This would constitute a clear case of biogeographic differentiation in large scale genealogical entities, a topic of interest to historical biogeographers. Further, many analyses have shown that there is an important relationship between times of major plate tectonic change and times of major evolutionary change (Hallam, 1967; Valentine and Moores, 1972; Valentine et al., 1978; Brooks and McLennan, 1991; Dalziel, 1997; Lieberman, 1997). This points out that there is some correlation between major evolutionary change and major biogeographic change.

Now, what about biogeographic patterns in entities of the economic hierarchy over these even longer time scales? Certainly plate tectonic events influence ecological biogeographic patterns in a region. For example, as a craton slowly drifts away from the equator, its diversity will change as a function of ecological biogeographic principles like latitudinal diversity gradients. However, large scale economic entities like communities and ecosystems seem to be largely ephemeral over time scales of ten's or hundreds of thousands of years, and therefore one might predict that these entities, and other even larger scale entities in the economic hierarchy like regional biotas, should be ephemeral over time scales of millions of years. Thus far though, this has not been conclusively demonstrated. In fact, it is a topic that is being actively debated by paleoecologists. Jackson (1992), Morris et al. (1995), Jackson et al. (1996), and Pandolfi (1996) have argued that although over time scales equivalent to those of Milankovitch cycles, communities and ecosystems are not persistent, over even longer time scales patterns of stability and persistence emerge in communities, ecosystems, and regional biotas.

Coordinated Stasis and Ecological Biogeographic Patterns

These hypothesized patterns of long term stability in large scale entities of the economic hierarchy have been referred to as coordinated stasis by Brett and Baird (1995). Coordinated stasis is a pattern recognized in fossiliferous Paleozoic rocks from eastern North America. During the Paleozoic much of eastern North America was a shallow seaway populated by invertebrate animals like rugose corals, brachiopods, and trilobites. In the rocks of this region, Brett and Baird (1995) documented the existence of faunal packages, which were 5–10 million year intervals of time that contained a large number of essentially unique species. The fauna in any given block shared few species with either the overlying or the underlying faunal package, such that faunal boundaries appeared to represent episodes of origination and extinction. Brett and Baird (1995) further suggested that during the existence of a fauna, species and communities were basically invariant. Building on the pattern described in Brett and Baird (1995), Morris et al. (1995) offered a mechanism to explain coordinated stasis: ecological locking. This mechanism involved a fauna's ability to withstand invasion by taxa from outside their region, and also to resist environmental perturbations.

If coordinated stasis and ecological locking are valid, this has great relevance to biogeography because then higher level ecological entities like communities and regional biotas would be stable through time due to the action of ecological mechanisms. Then long term, large scale ecological biogeographic patterns that could be studied in these faunas might include an association between tectonic events and patterns of faunal change, or how new faunas are assembled biogeographically.

Although provocative from the perspective of ecological biogeography, thus far coordinated stasis has not met with universal acceptance. Several authors, including Buzas and Culver (1994), Bennington and Bambach (1996), Holterhoff (1996), Jablonski and Sepkoski (1996), Westrop (1996), and Patzkowsky and Holland (1997) have argued that coordinated stasis is not a universal pattern in the fossil record, concluding that over long time scales and across broad spatial scales, entities like communities, ecosystems, and regional biotas are not cohesive and stable.

If coordinated stasis is ultimately rejected, then biogeographic patterns in higher level entities of the economic hierarchy will not really exist, or at least they will not be analogous to biogeographic patterns in higher level entities of the genealogical hierarchy. This would imply a significant disjunction between our ability to study biogeographic patterns in higher level entities of the genealogical and economic hierarchies. In particular, some clades might persist through long term tectonic changes such that the effects of geographic changes on these historical entities could be studied. By contrast, individual communities, ecosystems, and regional biotas will break apart in the face of profound tectonic changes, making it impossible to monitor biogeographic changes within any one of these types of ecological entities over long time scales.

If coordinated stasis is corroborated by future studies, then biogeographic patterns within communities, ecosystems, and regional biotas could be monitored over long periods of time, and entities from both hierarchies could be studied using partly congruent approaches. Further, if ecological locking is the mechanism that produces this pattern of coordinated stasis, then stability in regional biotas would be governed by ecological processes. This would be an important validation of a research program that considers biogeographic patterns in entities of the economic hierarchy, and also an important link between the different researchers studying biogeographic patterns in entities of the genealogical and economic hierarchies.

THE RELEVANCE OF RANGE EXPANSION TO HISTORICAL BIOGEOGRAPHY

Additional potential for links between biogeographic patterns in entities of the economic and genealogical hierarchies exists. Some of this potential revolves around the role of dispersal as a biogeographic process. Ecologists have long appreciated that dispersal is an important biogeographic process that influences biogeographic patterns (MacArthur and Wilson, 1967; Brown and Lomolino, 1998). However, what role should we assign to dispersal or processes like dispersal in influencing historical biogeographic patterns?

In historical biogeography, the term dispersal (or jump dispersal [Brundin, 1988]) has traditionally been defined as the expansion of a taxon's range over a geographic barrier with concomitant diversification (Humphries and Parenti, 1986; Nelson and Platnick, 1981; and others). It has been repeatedly demonstrated that dispersal, as traditionally defined, has little validity as a general explanation of historical biogeographic patterns for several reasons. One reason is that congruent patterns of biogeographic differentiation in different clades are unlikely to be the byproduct of this type of range expansion. When this phenomenon occurs, biogeographic patterns will instead depend on the individual ecological characteristics of taxa (Croizat et al., 1974; Nelson and Platnick, 1978, 1981; Platnick and Nelson, 1978; Rosen, 1978, 1979; Brooks, 1981; Wiley, 1981, 1988a, b; Kluge, 1988; Brooks and McLennan, 1991). Another reason is that historical biogeographic explanations that invoke this type of range expansion often deteriorate into a series of narratives. Finally, these types of long-distance dispersal events are probably quite rare. Traditional dispersal, however, is not the only type of range expansion that has been postulated to occur by phylogenetic systematists. There are other types of range expansion, and some of these produce congruent historical biogeographic patterns.

Other Types of Range Expansion

One additional type of range expansion that has been defined is dispersion (Platnick, 1976). This involves the expansion or diffusion of a population or species over a homogeneous landscape until it encounters geographic barriers. This type of range expansion as originally defined is unlike traditional dispersal in that it is not associated with diversification. That is, there are no cladogenetic events that occur in such dispersion.

Rather, it occurs within a single taxon. Thus, dispersion is a type of range expansion acknowledged to occur, yet it is not related to cladogenetic diversification, nor is it necessarily an event that needs to occur congruently in several different taxa. Thus, it is unlikely to find explanation within a historical biogeographic framework.

Congruent patterns of range expansion that would be recognized by biogeographers must involve cladogenesis. Principally, a geographic barrier, either geological or climatic, needs to fall such that several taxa can simultaneously expand their ranges via dispersion (Figs. 3A, B). Then, in many clades, the taxa will go from being present in area A to present in area A and area B, though really areas A and B have now combined into a larger area. If such an event occurs it will be congruent in the sense that it occurs in several independent taxa, but it does not involve cladogenesis. Thus, it is not immediately relevant to historical biogeography.

For this type of range expansion to be relevant to historical biogeography a second ingredient is needed. What is required is that later the previous barrier, or perhaps another barrier within the newly expanded ranges, rises again, producing allopatric speciation and vicariance in several taxa (Fig. 3C). This type of biogeographic pattern has been uncovered by

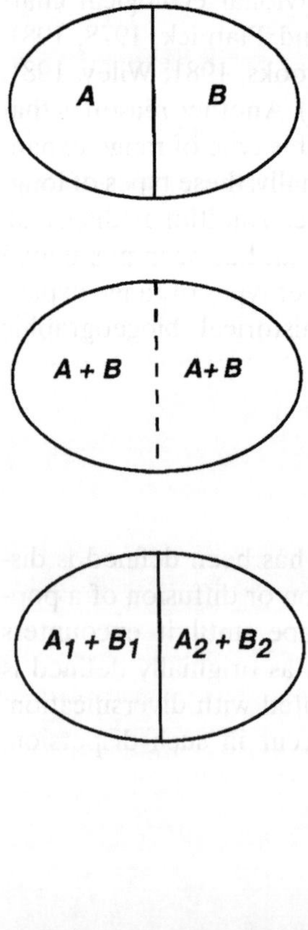

FIG. 3. A hypothetical example of congruent range expansion (illustrated in a single taxon for the purposes of clarity): A) A hypothetical continent which is bisected by a geographic barrier. Prior to the emergence of this barrier there was a single ancestral species distributed across the entire continent. When the barrier emerged, populations became isolated on either side of the barrier, differentiated and then speciated via vicariance such that there were two closely related species separated by a geographic barrier. B) At a later time, the geographic barrier was removed such that both species A and B could expand their range into areas they previously did not occupy. C) Still later, a geographic barrier re-emerges. Now populations of both species A and species B are isolated on either side of the barrier. They become isolated, differentiate and eventually speciate such that there are now a total of four species, two on each side of the geographic barrier.

Lieberman and Eldredge (1996) from the marine invertebrate fossil record during the time of tectonic collision between North America and Europe in the Middle Devonian 380 million years ago, which led to the initial uplift of the Appalachian mountain chain. Several clades of marine invertebrates that occupied North America showed not only congruent episodes of vicariant differentiation, but also episodes of congruent range expansion followed by subsequent vicariance. This congruent range expansion was referred to as geo-dispersal by Lieberman and Eldredge (1996) to convey that the episodes of range expansion were mediated by geological events, specifically tectonic collisions that brought isolated faunas together. These events are not equivalent to traditional dispersal, which involves range expansion in which organisms extend their ranges across a barrier. Instead, geo-dispersal involves cyclical geological events, which first eliminate barriers and allow range expansion in several taxa, and then cause the emergence of the same or a new barrier which leads to subsequent vicariance.

Historical Perspective On Geo-Dispersal

This phenomenon has been recognized by other phylogenetic systematists as well. For example, Platnick and Nelson (1978, p. 7) invoked a specific example, the tectonic collision between India and Asia during the Cenozoic Era, with the subsequent range expansion of Indian taxa into Asia and Asian taxa into India. Brundin (1988, p. 348 and p. 362–3) suggested that geographic barriers may at various times disappear due to the vicissitudes of geological change. When this happens, range expansion in several taxa results in a manner analogous to geo-dispersal. "Vicariance and range expansion have constantly alternated, bringing about intermittent interchange in biotas of different parts of the world" (Brundin, 1988, p. 356–7). An actual clade that possibly displays such a pattern is shown in Fig. 4. Cracraft (1988, p. 233) and Noonan (1988, p. 377) discussed how geographic barriers prohibiting the free movement of taxa oscillate states between present and absent, such that there could be vicariance and subsequent range expansion (i.e., geo-dispersal). Wiley (1988a, p. 271) recognized that when geographic barriers fall entire biotas can expand their range. Ronquist (1998, p. 320) has also advocated the idea that there is a difference between range expansion produced when geographic barriers fall (geo-dispersal, though he did not call it that) and traditional dispersal. Finally, Bremer (1992), Hovenkamp (1997), and Ronquist (1997) have all discussed in general terms both the potential for range expansion due to the fall of geographic barriers and the prevalence of reticulation or joining of areas.

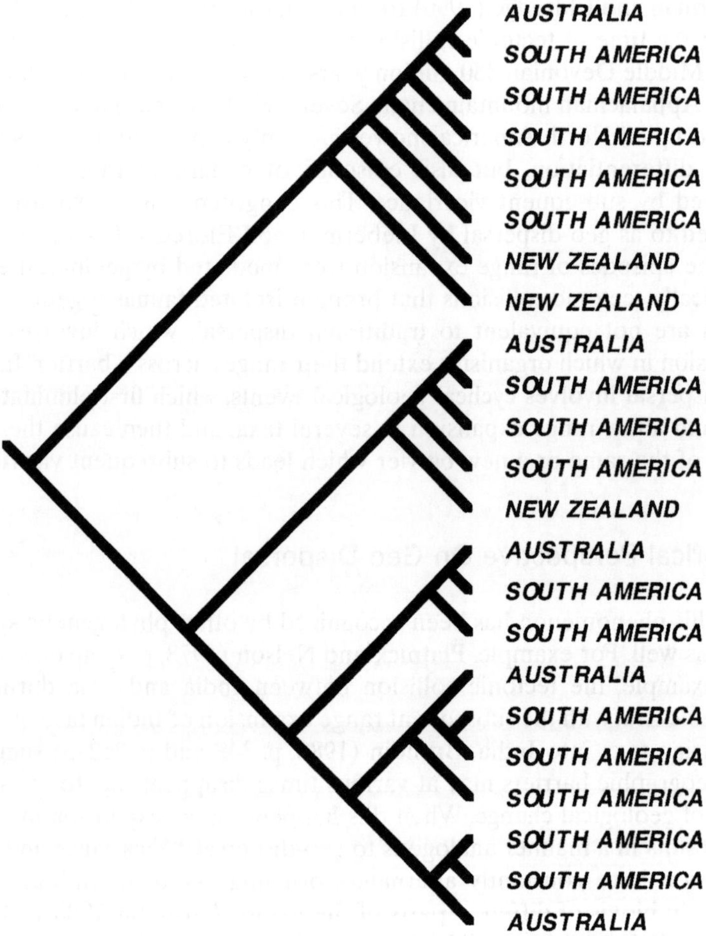

FIG. 4. An area cladogram modified from Brundin (1988) which shows an actual example, based on a phylogeny of insects, that provides evidence within a single clade for multiple episodes of vicariance, followed by range expansion, followed by subsequent vicariance.

Although geo-dispersal is an idea that has been previously postulated, there has not been a method proposed to distinguish between it and vicariance before that of Lieberman and Eldredge (1996). It turns out that geo-dispersal can be analyzed within a historical biogeographic framework using the two primary methods of biogeographic analysis, Component Analysis and Brooks Parsimony Analysis (BPA) (Lieberman, 2000).

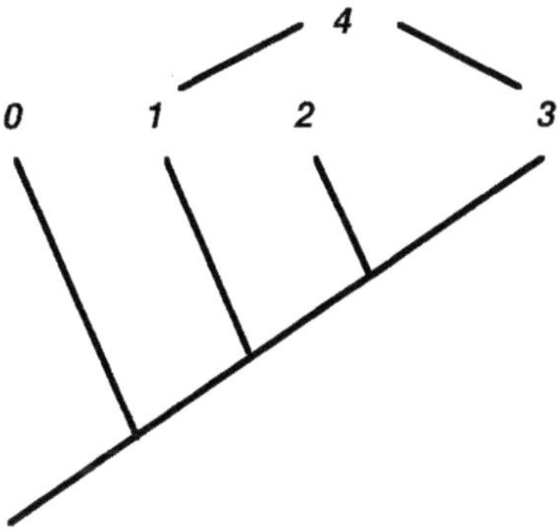

FIG. 5. An area cladogram adapted from Rosen (1978), where there is evidence for an area hybridization event between areas 1 and 3 in area 4.

Analytical Approaches To Study Geo-Dispersal

In Component Analysis, geo-dispersal could be identified using the concept of area hybridization discussed by Rosen (1978) (Fig. 5). If similar evidence for hybridization between the same two or more areas was documented in several clades it would provide evidence that geographic barriers formerly separating these areas had fallen, allowing several taxa to expand concurrently their ranges (see Fig. 6).

In BPA, geo-dispersal could also be documented using the parsimony-based matrix approach on which this method relies (BPA was described by Brooks [1985, 1990], Brooks and McLennan [1991], Wiley [1988a, b], and Wiley et al. [1991]). Episodes of range expansion could be coded into a separate BPA matrix that could be analyzed to search for geo-dispersal in a manner analogous to the typical BPA that searches for episodes of vicariance (Fig. 6; Table I). (Protocols for coding a geo-dispersal matrix are discussed more fully by Lieberman and Eldredge [1996] and Lieberman [1997].) If similar patterns of range expansion were replicated in several

FIG. 6. An area cladogram derived from a hypothetical four taxon cladogram illustrating how BPA can be used to study congruent episodes of range expansion or geo-dispersal (also see Table I). T1–T4 refers to terminal taxa 1–4. One of the taxa (T4) was widespread in areas 1, 2, and 3. The geographic distribution of the ancestral nodes (N1–N3) was inferred using Fitch (1971) optimization. (Any type of optimization could be used, see discussion in Bremer [1995] and Ronquist, [1994].)

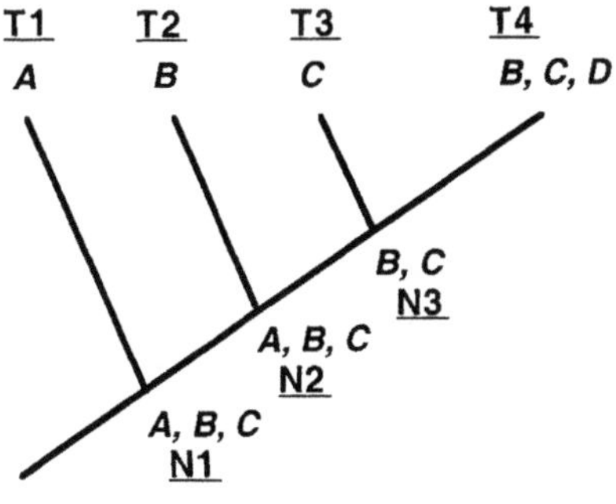

TABLE I. BPA matrix coding for analysis of geo-dispersal based on the area cladogram of Fig. 4. In standard BPA matrix coding there are potentially seven characters (3 nodes labeled N1–N3 and 4 terminal taxa labeled T1–T4 in Fig. 4). However, only one of these 7 potential characters, that describing the transition between N3 and T4, provides evidence for range expansion. Thus, it would be the only one coded in this variant of BPA. "0" is treated as the primitive condition, absent from an area; "1" implies presence in an area; "2" implies range expansion into a new area relative to an immediate ancestor. This character would be treated as additive (ordered), to provide information about the direction of range expansion. 'ANCESTOR' is the standard outgroup typically employed in BPA.

TAXA (AREAS)	CHARACTER T4
ANCESTOR	0
A	0
B	1
C	1
D	2

different clades, this should emerge in an area cladogram produced by parsimony analysis of a BPA geo-dispersal matrix. If these types of range expansion are common, such geo-dispersal area cladograms should be well resolved and be strongly supported by various indices of tree support. By contrast, if such types of range expansion are rare or infrequent, geo-dispersal area cladograms should show little resolution or be poorly supported according to various measures of tree support. The time period (present versus deep past) and the region being studied are likely to play a role in determining the prevalence of geo-dispersal (Lieberman and Eldredge, 1996; Lieberman, 1997). At times in earth history some regions have been influenced powerfully by such geological process as continental rifting, which cause vicariance. At other times and in other regions processes that might have contributed to geo-dispersal may have prevailed, such as the collision of Asia and India (Platnick and Nelson, 1978). Studies can quantify to what extent this may be a more or less important biogeographic process (Lieberman and Eldredge, 1996; Lieberman, 1997).

There has been extensive debate about which of the competing biogeographic approaches described above, Component Analysis or BPA, is most effective for resolving biogeographic patterns. Methodological issues and ontological/philosophical issues are involved in this debate (see Wiley, 1988a, b; Brooks, 1990; Brooks and McLennan, 1991; Morrone and Carpenter, 1994; Morrone and Crisci, 1995; Kluge, 1997, 1998; and Van Soest and Hajdu, 1997). Although the debate has not yet been resolved, compelling evidence seems to be emerging to suggest that BPA is equally or more effective as an exploratory tool for retrieving biogeographic patterns in general, and patterns of geo-dispersal in particular (Lieberman, 2000). Further, BPA appears to have a more rigorous ontological underpinning.

CONCLUSIONS

Although traditionally ecological and historical biogeography have been treated as distinct subdisciplines, with little prospect for interdigitation, evidence is mounting to suggest that this no longer needs to be the case. For example, extrapolating the principle that the natural world is hierarchically arrayed to the subject of biogeography provides a clear-cut illustration of when and why ecological and historical biogeographic patterns and processes should correspond, and when and why they should diverge. They should correspond particularly at the population level, where there is the strong possibility of congruence between historical and ecological biogeographic differentiation, as exemplified by the phenomenon of coevolution. At larger scale entities in life's two hierarchies ecological and historical biogeographic patterns may diverge. However, biogeographic patterns may be equally important and worthy of study in both large scale entities of the economic and genealogical hierarchies. For example, historical biogeographic processes clearly have played a role in shaping the evolution and distribution of the largest scale entities in the genealogical hierarchy, and research from paleoecology suggests that ecological biogeographic processes may also play a role in shaping the distribution and ecology of the largest scale entities in the economic hierarchy. Moreover, range expansion was treated as a significant biogeographic process by ecological biogeographers only, whereas historical biogeographers viewed it as a kind of noise obscuring vicariance, the signal of interest. However, the existence of geo-dispersal as a congruent biogeographic phenomenon, and the demonstration that patterns of range expansion can be analyzed in exactly the same manner as patterns of vicariance, suggests that the disciplines of ecological and historical biogeography need not always be disjunct.

ACKNOWLEDGMENTS

The comments of D. Brooks, R. Kaesler, and E. O. Wiley on earlier versions of this manuscript are greatly appreciated. Thanks also to R.

MacIntyre. This research was supported by a NFGRF Grant from the University of Kansas, NSF grants EPS-9874732 and OPP-9909302.

REFERENCES

Allen, T. H. F., and Starr, T. B., 1982, *Hierarchy: Perspectives for Ecological Complexity.* University of Chicago Press, Chicago, IL.

Arnold, A. J., and Fristrup, K., 1982, The theory of evolution by natural selection: a hierarchical expansion. *Paleobiology* **8:**113–129.

Avise, J. C., 1986, Mitochondrial DNA and the evolutionary genetics of higher animals. *Phil. Trans. R. Soc. London, Ser. B* **312:**325–342.

Avise, J. C., 1992, Molecular population structure and the biogeographic history of a regional fauna: a case history with lessons for conservation biologyq. *Oikos* **63:**62–76.

Avise, J. C., Arnold, J., Ball, R. M., Bermingham, E., Lamb, T., Neigel, J. E., Reeb, C. A. and Saunders, N. C., 1987, Intraspecific phylogeography: the mitochondrial DNA birdge between population genetics and systematics. *Ann. Rev. Ecol. Syst.* **18:**489–522.

Bennett, K. D., 1990, Milankovitch cycles and their effects on species in ecological and evolutionary time. *Paleobiology* **16:**11–21.

Bennington, J. B., and Bambach, R. K., 1996, Statistical testing for paleocommunity recurrence: Are similar fossil assemblages ever the same? *Palaeogeog., Palaeoclim., and Palaeoec.* **127:**107–134.

Berger, A., 1980, The Milankovitch astronomical theory of paleoclimates: a modern review in: *Vistas on Astronomy* (A. Beer, K. Pounds, and P. Beer, eds.). pp. 103–122 Pergamon Press, London.

Bremer, K., 1992, Ancestral areas: a cladistic reinterpretation of the center of origin concept. *Syst. Biol.* **41:**436–445.

Bremer, K., 1995, Ancestral areas: optimization and probability. *Syst. Biol.* **44:**255–259.

Brett, C. E., and Baird, G. C., 1995, Coordinated stasis and evolutionary ecology of Silurian to Middle Devonian faunas in the Appalachian Basin in: *New approaches to speciation in the fossil record* (D. H. Erwin, and R. L. Anstey, eds.) pp. 285–315 Columbia University Press, New York, NY.

Brooks, D. R., 1981, Hennig's parasitological method: a proposed solution. *Syst. Zool.* **30:**229–249.

Brooks, D. R., 1985, Historical ecology: a new approach to studying the evolution of ecological associations. *Ann. Miss. Bot. Gard.* **72:**660–680.

Brooks, D. R., 1988, Scaling effects in historical biogeography: a new view of space, time, and form. *Syst. Zool.* **37:**237–244.

Brooks, D. R., 1990, Parsimony analysis in historical biogeography and coevolution: methodological and theoretical update. *Syst. Zool.* **39:**14–30.

Brooks, D. R., and McLennan, D. A., 1991, *Phylogeny, Ecology, and Behavior.* University of Chicago Press, Chicago.

Brooks, D. R., and Wiley, E. O., 1986, *Evolution as Entropy. Toward a Unified Theory of Biology.* University of Chicago Press, Chicago, IL.

Brown, J. H., and Lomolino, M. V., 1998, *Biogeography*, 2nd edition, Sinauer, Sunderland, MA.

Brown, J. H., and Maurer, B. A., 1989, Macroecology: the division of food and space among species on continents. *Science* **243:**1143–1150.

Browne, J., 1983, *The Secular Ark: Studies in the History of Biogeography*. Yale University Press, New Haven, CT.

Brundin, L. Z., 1988, Phylogenetic biogeography in: *Analytical Biogeography* (A. A. Myers, and P. S. Giller, eds.) pp. 343–369 Chapman and Hall, New York City.

Burns, T. P., Patten, B. C., and Higashi, M., 1991, Hierarchical evolution in ecological networks: environs and selection in: *Theoretical studies of ecosystems: the network perspective* (M. Higashi, and T. P. Burns, eds.) pp. 211–239 Cambridge University Press, New York.

Buss, L. W., 1987, *The Evolution of Individuality*. Princeton University Press, Princeton, NJ.

Buzas, M. A., and Culver, S. J., 1994, Species pool and dynamics of marine paleocommunities. *Science* **264:**1439–1441.

Coope, G. R., 1979, Late Cenozoic Coleoptera: evolution, biogeography, and ecology. *Ann. Rev. Ecol. Syst.* **10:**247–267.

Coope, G. R., 1990, The invasion of northern Europe during the Pleistocene by Mediterranean species of Coleoptera in: Biological Invasions in Europe and the Mediterranean basin (F. d. Castri, A. J. Hansen, and M. Debussche, eds.) pp. 203–215 Kluwer Academic Publishers, Dordrecht, Netherlands.

Cracraft, J., 1982, Geographic differentiation, cladistics, and vicariance biogeography: reconstructing the tempo and mode of evolution. *Amer. Zool.* **22:**411–424.

Cracraft, J., 1988, Deep-history biogeography: retrieving the historical pattern of evolving continental biotas. *Syst. Zool.* **37:**221–236.

Croizat, L., Nelson, G., and Rosen, D. E., 1974, Centers of origin and related concept. *Syst. Zool.* **23:**265–287.

Dalziel, I. W. D., 1997, Neoproterozoic-Paleozoic geography and tectonics: review, hypothesis, and environmental speculations. *Geol. Soc. Am. Bull.* **109:**16–42.

Damuth, J., 1985, Selection among "species": a formulation in terms of natural functional units. *Evolution* **39:**1132–1146.

Davis, M. B., 1976, Pleistocene biogeography of temperate deciduous forests. *Geosci. Man* **13:**13

Davis, M. B., 1986, Climatic instability, time lags, and community disequilibrium in: *Community Ecology* (J. Diamond, and T. Case, eds.) pp. 269–284 Harper and Row, New York.

Dawkins, R., 1982, *The Extended Phenotype*. W. H. Freeman, San Francisco, CA.

Dobzhansky, T., 1937, *Genetics and the Origin of Species*. Columbia University Press, New York.

Eldredge, N., 1982, Phenomenological levels and evolutionary rates. *Syst. Zool.* **31:**338–347.

Eldredge, N., 1985, *Unfinished Synthesis*. Oxford University Press, New York.

Eldredge, N., 1989, *Macroevolutionary Dynamics*. McGraw Hill, New York.

Eldredge, N., 1995, *Reinventing Darwin*. John Wiley & Sons, New York.

Eldredge, N., and Cracraft, J., 1980, *Phylogenetic Patterns and the Evolutionary Process: Method and Theory in Comparative Biology*. Columbia University Press, New York.

Eldredge, N., and Ormiston, A. R., 1979, Biogeography of Silurian and Devonian trilobites of the Malvinokaffric Realm in: *Historical Biogeography, Plate Tectonics, and the Changing Environment* (J. Gray, and A. J. Boucto, eds.) pp. 147–167 Oregon State University, Corvallis, OR.

Eldredge, N., and Salthe, S. N., 1984, Hierarchy and evolution. *Oxf. Surv. Evol. Biol.* **1:**184–208.

Feynman, R., 1965, *The Character of Physical Law*. MIT Press, Cambridge, MA.

Fitch, W. M., 1971, Toward defining the course of evolution: minimum change for a specific tree topology. *Syst. Zool.* **20:**406–416.

Foster, D. R., Schoonmaker, P. K., and Pickett, S. T. A., 1990, Insights from paleoecology to community ecology. *Tr. Ecol. Evol.* **5:**119–122.

Funk, V. A., and Brooks, D. R., 1990, Phylogenetic systematics as the basis of comparative biology. *Smithsonian Contr. Bot.* **73:**1–45.

Goodwin, B., 1994, *How the Leopard Changed its Spots*. Charles Scribner's Sons, New York.

Gould, S. J., 1980, Is a new and general theory of evolution emerging? *Paleobiology* **6:**119–130.

Gould, S. J., 1982, Darwinism and the expansion of evolutionary theory. *Science* **216:**380–387.

Gould, S. J., 1990, Speciation and sorting as the source of evolutionary trends, or "things are seldom what they seem," in: *Evolutionary Trends* (K. J. McNamara, ed.) pp. 3–27 Belhaven Press, London.

Graham, R. W., 1986, Response of mammalian communities to environmental changes during the late Quaternary in: *Community Ecology* (J. Diamond, and T. J. Case, eds.) pp. 300–313 Harper and Row, New York.

Graham, R. W., 1992, Late Pleistocene faunal changes as a guide to understanding effects of greenhous warming on the mammalian fauna of North America in: *Global Warming and Biological Diversity* (R. L. Peters, and T. E. Lovejoy, eds.) pp. 76–87 Yale University Press, New Haven, CT.

Graham, R. W., Lundelius, E. L., Jr., Graham, M. A., Schroeder, E. K., Toomey, R. S., III, Anderson, E., Barnosky, A. D., Burns, J. A., Churcher, C. S., Grayson, D. K., Guthrie, R. D., Harington, C. R., Jefferson, G. T., Martin, L. D., McDonald, H. G., Morlan, R. E., Semkin, H. A., Jr., Webb, S. D., Werdelin, L., and Wilson, M. C., 1996, Spatial response of mammals to late Quaternary environmental fluctuations. *Science* **272:**1601–1606.

Hallam, A., 1967, The bearing of certain paleogeographic data on continental drift. *Palaeogeog., Palaeoclim., and Palaeoec.* **3:**201–224.

Hays, J. D., Imbrie, J., and Shackleton, N. J., 1976, Variations in the Earth's orbit:pacemaker of the ice ages. *Science* **194:**1121–1132.

Holterhoff, P., 1996, Crinoid biofacies in Upper Carboniferous cyclothems, midcontinent North America: faunal tracking and the role of regional processes in biofacies recurrence. *Palaeogeog., Palaeoclim., and Palaeoec.* **127:**47–82.

Hovenkamp, P., 1997, Vicariance events, not areas, should be used in biogeographical analysis. *Cladistics* **13:**67–79.

Hull, D. L., 1988, *Science as a Process*. University of Chicago Press, Chicago, IL.

Hull, D. S., 1980, Individuality and selection. *Ann. Rev. Ecol. Syst.* **11:**311–332.

Humphries, C. J., and Parenti, L., 1986, Cladistic biogeography. *Oxf. Monog. Biogeog.* **2:**1–98.

Huntley, B., and Webb, T., III, 1989, Migration: species' respnse to climatic variations caused by changes in the earth's orbit. *J. Biogeog.* **16:**5–19.

Imbrie, J., and Imbrie, J. Z., 1980, Modeling the climatic response to orbital variations. *Science* **207:**943–953.

Jablonski, D., and Sepkoski, J. J., Jr., 1996, Paleobiology, community ecology, and scales of ecological pattern. *Ecology* **77:**1367–1378.

Jackson, J. B. C., 1992, Pleistocene perspectives on coral reef community structure. *Amer. Zool.* **32:**719–730.

Jackson, J. B. C., Budd, A. F., and Pandolfi, J. M., 1996, The shifting balance of natural communities? in: *Evolutionary Paleobiology* (D. Jablonski, D. H. Erwin, and J. H. Lipps, eds.) pp. 203–220 University of Chicago Press, Chicago, IL.

Kluge, A. G., 1988, Parsimony in vicariance biogeography: a quantitative method and a Greater Antillean example. *Syst. Zool.* **37:**315–328.

Kluge, A. G., 1997, Testability and the refutation and corroboration of cladistic hypotheses. *Cladistics* **13:**81–96.

Kluge, A. G., 1998, Total evidence or taxonomic congruence: cladistics or consensus classification. *Cladistics* **14:**151–158.

Lieberman, B. S., 1995, Phylogenetic trends and speciation: analyzing macroevolutionary processes and levels of selection in: *New Approaches to Speciation in the Fossil Record* (D. H. Erwin, and R. L. Anstey, eds.) pp. 316–337 Columbia University Press, New York.

Lieberman, B. S., 1997, Early Cambrian paleogeography and tectonic history: a biogeographic approach. *Geology* **25:**1039–1042.

Lieberman, B. S., 2000, *Paleobiogeography: Using Fossils to Study Global Change, Plate Tecontics, and Evolution*. Kluwer Academic Press/Plenum Publishing, New York.

Lieberman, B. S., Allmon, W. D., and Eldredge, N., 1993, Levels of selection and macroevolutionary patterns in the turritellid gastropods. *Paleobiology* **19:**205–215.

Lieberman, B. S., and Eldredge, N., 1996, Trilobite biogeography in the Middle Devonian: geological processes and analytical methods. *Paleobiology* **22:**66–79.

Lieberman, B. S., and Vrba, E. S., 1995, Hierarchy theory, selection, and sorting: a phylogenetic perspective. *BioScience* **45:**394–399.

MacArthur, R. H., and Wilson, E. O., 1967, *The Theory of Island Biogeography*. Princeton University Press, Princeton, NJ.

Mayden, R. L., 1988, Vicariance biogeography, parsimony, and evolution in North American freshwater fishes. *Syst. Zool.* **37:**329–355.

Mayr, E., 1942, *Systematics and the Origin of Species*. Dover Press, New York, NY.

Mayr, E., 1982, *The Growth of Biological Thought*. Harvard University Press, Cambridge, MA.

McGowan, J. A., Cayan, D. R., and Dorman, L. M., 1998, Climate-ocean variability and ecosystem response in the northeast Pacific. *Science* **281:**210–217.

Morris, P. J., Ivany, L. C., Schopf, K. M., and Brett, C. E., 1995, The challenge of paleoecological stasis: reassessing sources of evolutionary stability. *Proc. Nat. Acad. Sci., USA* **92:**11269–11273.

Morrone, J. J., and Carpenter, J. M., 1994, In search of a method for cladistic biogeography: an empirical comparison of Component Analysis, Brooks Parsimony Analysis, and Three-Area statements. *Cladistics* **10:**99–153.

Morrone, J. J., and Criscis, J. V., 1995, Historical Biogeography: introduction to methods. *Ann. Rev. Ecol. Syst.* **26:**373–401.

Nelson, G., 1978, From Candolle to Croizat: comments on the history of biogeography. *J. Hist. Biol.* **11:**269–305.

Nelson, G., and Platnick, N. I., 1978, The perils of plesiomorphy: widespread taxa, dispersal and phenetic biogeography. *Syst. Zool.* **27:**474–477.

Nelson, G., and Platnick, N. I., 1981, *Systematics and Biogeography: Cladistics and Vicariance*. Columbia University Press, New York City.

Noonan, G. R., 1988, Biogeography of North American and Mexican insects, and a critique of vicariance biogeography. *Syst. Zool.* **37:**366–384.

Pandolfi, J. M., 1996, Limited membership in Pleistocene reef coral assemblages from the Huon Peninsula, Papua New Guinea: constancy during global change. *Paleobiology* **22:**152–176.

Patzkowsky, M. E., and Holland, S. M., 1997, Patterns of turnover in Middle and Upper Ordovician brachiopods of the eastern United States: a test of coordinated stasis. *Paleobiology* **23:**420–443.

Platnick, N. I., 1976, Concepts of dispersal in historical biogeography. *Syst. Zool.* **25:**294–295.

Platnick, N. I., and Nelson, G., 1978, A method of analysis for historical biogeography. *Syst. Zool.* **27:**1–16.

Ricklefs, R. E., 1987, Community diversity: relative roles of local and regional processes. *Science* **253:**167–171.

Ronquist, F.,1994, Ancestral areas and parsimony. *Syst. Biol.* **43:**267–274.

Ronquist, F., 1997, Dispersal-vicariance analysis: a new approach to the quantification of historical biogeography. *Syst. Biol.* **46:**195–203.

Ronquist, F., 1998, Phylogenetic approaches in coevolution and biogeography. *Zool. Scr.* **26:**313–322.

Rosen, B., 1991, *Life Itself*. Columbia University Press, New York.

Rosen, D. E., 1978, Vicariant patterns and historical explanation in biogeography. *Syst. Zool.* **27:**159–188.

Rosen, D. E., 1979, Fishes from the uplands and intermontane basins of Guatemala: revisionary studies and comparative geography. *Bull. Amer. Mus. Nat. Hist.* **162:**269–375.

Ross, H. H., 1972, The origin of species diversity in ecological communities. *Taxon* **21:**253–259.

Ross, H. H., 1986, Resource partitioning in fish assemblages: a review of field studies. *Copeia* **86:**352–388.

Salthe, S. N., 1985, *Evolving Hierarchical Systems*. Columbia University Press, New York.

Smuts, J. C., 1925, *Holism and Evolution*. Viking Press, New York.

Sober, E., 1984, *The Nature of Selection*. MIT Press, Cambridge, MA.

Stevens, G., 1992, Spilling over the competitive limits to species coexistence in: *Systematics, Ecology, and the Biodiversity Crisis* (N. Eldredge, ed.) pp. 40–58 Columbia University Press, New York.

Streidter, G. F., and Northcutt, R. G., 1991, Biological hierarchies and the concept of homology. *Brain Behav. Evol.* **38:**177–189.

Valentine, J. W., Foin, T. C., and Peart, D., 1978, A provincial model of Phanerozoic marine diversity. *Paleobiology* **4:**55–66.

Valentine, J. W., and May, C. L., 1996, Hierarchies in biology and paleontology. *Paleobiology* **22:**23–33.

Valentine, J. W., and Moores, E. M., 1972, Global tectonics and the fossil record. *J. Geol.* **80:**167–184.

Van Soest, R. W. M., and Hajdu, E., 1997, Marine area relationships from twenty sponge phylogenies. A comparison of methods and coding strategies. *Cladistics* **13:**1–20.

Vermeij, G., 1978, *Biogeography and Adaptation*. Harvard University Press, Cambridge, MA.

Vrba, E. S., 1980, Evolution, species and fossils: how does life evolve? *S. Afr. J. Sci.* **76:**61–84.

Vrba, E. S., 1983, Macroevolutionary trends: new perspectives on the roles of adaptation and incidental effect. *Science* **221:**387–389.

Vrba, E. S., 1985, Environment and evolution: alternative causes of the temporal distribution of evolutionary events. *S. Afr. J. Sci.* **81:**229–236.

Vrba, E. S., 1987, Ecology in relation to speciation rates: some case histories of Miocene-Recent mammal clades. *Evol. Ecol.* **1:**283–300.

Vrba, E. S., 1989, Levels of selection and sorting with special reference to the species level. *Oxf. Surv. Evol. Biol.* **6:**111–168.

Vrba, E. S., 1993, Turnover-pulses, the Red Queen, and related topics. *Am. J. Sci.* **293:**418–452.

Vrba, E. S., 1996, On the connection between paleoclimate and evolution in: *Paleoclimate and Evolution with Emphasis on Human Origins* (E. S. Vrba, G. H. Denton, T. C. Partridge, and L. H. Burckle, eds.) pp. 24–45 Yale University Press, New Haven, CT.

Vrba, E. S., and Eldredge, N., 1984, Individuals, hierarchies and processes: toward a more complete evolutionary theory. *Paleobiology* **10:**146–171.

Vrba, E. S., and Gould, S. J., 1986, The hierarchical expansion of sorting and selection: sorting and selection cannot be equated. *Paleobiology* **12:**217–228.

Westrop, S. R., 1996, Temporal persistence and stability of Cambrian biofacies: Sunwaptan (Upper Cambrian) trilobite faunas of North America. *Palaeogeog. Palaeoclimat., and Palaeoec.* **127:**33–46.

Wiley, E. O., 1981, *Phylogenetics: The Theory and Practice of Phylogenetic Systematics*. Wiley, New York City.

Wiley, E. O., 1986, Methods in vicariance biogeography in: *Systematics and Evolution: a matter of diversity* (P. Hovenkamp ed.) pp. 283–306 University of Utrecht Press, Utrecht, The Netherlands.

Wiley, E. O., 1988a, Parsimony analysis and vicariance biogeography. *Syst. Zool.* **37:**271–290.

Wiley, E. O., 1988b, Vicariance biogeography. *Ann. Rev. Ecol. Syst.* **19:**513–542.

Wiley, E. O., and Mayden, R. L., 1985, Species and speciation in phylogenetic systematics, with examples from the North American fish fauna. *Ann. Miss. Bot. Gard.* **72:**596–635.

Wiley, E. O., Siegel-Causey, D., Brooks, D. R., and Funk, V. A., 1991, *The Compleat Cladist.* University of Kansas Press, Lawrence, Kansas.

Williams, G. C., 1992, *Natural Selection*. Oxford University Press, New York.

Zink, R. M., 1991, The geography of mitochondrial DNA variation in two sympatric sparrows. *Evolution* **45:**329–339.

2

The Evolutionary Origin and Maintenance of Sexual Recombination: A Review of Contemporary Models

JOHN A. BIRDSELL and CHRISTOPHER WILLS

INTRODUCTION

This review is intended to be a global examination of the various hypotheses for the origin and maintenance of genetic recombination and outcrossing, with a look at the surprisingly limited amounts of experimental evidence that has been obtained in order to distinguish among them. It is designed for the reader who wishes an overview of the current state of this large and complex field. However, it is simply not possible to deal in detail with the many competing and complementary hypotheses without turning the review into a book. We apologize in advance to those whose ideas and contributions may have been left out, or may not have been dealt with in the detail that they would like.

For readers interested in other detailed examinations of current theories, the following references should be consulted: Bell (1982); Bernstein and Bernstein (1991); Ghiselin (1974); Maynard Smith (1978);

JOHN A. BIRDSELL • Department of Ecology and Evolutionary Biology, University of Arizona, Tucson, AZ 85721. CHRISTOPHER WILLS • Department of Biology and Center for Molecular Genetics, University of California, San Diego, La Jolla, CA 92093-0116.

Evolutionary Biology, Volume 33
Edited by MacIntyre and Clegg, Kluwer Academic/Plenum Publishers, 2003

Margulis and Sagan (1985); Michod and Levin (1988); Stearns (1987); Williams (1975).

When discussing any topic as complicated as the evolution of sex, it is useful to have working definitions of the terms used by various workers in the field. To aid with this, a glossary of terms has been compiled from Margulis and Sagan (1984) and Margulis and Sagan (1985) as well as a number of other sources.

Definitions of Sex

The term "sex" has different meanings to different people. Vernacular usage aside, the term implies "reproduction" or "procreation" to most non-evolutionary biologists. This need not be the case, however, for sex can, in fact, be completely decoupled from reproduction, as is the case in the yeast *Saccharomyces cerevisiae*. This simple single-celled eukaryote reproduces mitotically by budding, in which a "mother" cell gives rise to a genetically identical (but smaller) daughter cell (Herskowitz, 1988). Sex in this organism only occurs when two cells of opposite mating types come into contact and fuse to produce a diploid cell incapable of any further mating. Two things can then happen to this diploid: (1) it can, under stressful conditions, be induced to undergo meiosis and recombination to produce four genetically different haploid cells, thereby completing the sexual cycle, or (2) it can stably reproduce by mitosis (Herskowitz, 1988).

Even among evolutionary biologists there is no consensus as to how to define sex. Seger and Hamilton (1988) fail to encompass the type of sex just described in their use of the term, which to them means "nominally biparental reproduction involving differentiated male and female individuals or reproductive functions." According to Bell (1982), sex is "a composite process in the course of which genomes are diversified by a type of nuclear division called meiosis, and by a type of nuclear fusion called syngamy, or fertilization." Of course this definition leaves out the occurrence of recombination mechanisms in non-eukaryotic organisms such as bacteria and viruses.

Margulis et al. (1985) feel that "bacterial sex" should be included in the definition of sex. They prefer to define sex as "a process always involving at least one autopoietic parent that forms an individual with a genetic constitution that differs from both (sic) of the parents." Autopoiesis is derived from the Greek for "self-making," and refers to an organism's ability to maintain itself metabolically by utilizing carbon and energy resources (Margulis and Sagan, 1984; Margulis et al., 1985). Thus, according to Margulis et al., the term "sex" includes processes occurring in both eukaryotes as well as prokaryotes. It does not, however, include processes that take place in viruses. They also point out the importance of distinguishing between prokaryotic sex (conjugation, transformation, and transduction) and eukaryotic (meiotic) sex.

The fact that sex-like activities do occur in viruses (multiplicity reactivation) has led Bernstein et al. (1984) to the broadest definition of sex, which they define as "a process in which the genomes of two parents are brought together in a common cytoplasm to produce progeny which may then contain reassorted portions of the parental genomes." It is this latter, broadest definition of sex which will be used throughout this review.

As can be seen from the numerous ways in which sex can be defined, it is very important to state clearly what type of sex is referred to when attempting to discuss its evolution.

"Recombination" is another term intimately associated with sex, which again has caused much confusion and controversy due to the fact that there is no universally accepted usage of the term. To many, recombination is synonymous with "mixis" and the production of new combinations of pre-existing alleles at many homologous loci (Michod and Levin, 1988). Such new combinations can be brought about by a number of means acting on several different levels. Physical recombination (breakage and reunion of DNA molecules) that takes place between homologous chromosomes can result in new combinations of pre-existing alleles. However, this need not be the case. It is important to distinguish between physical recombination (the breakage and reunion of DNA molecules), and its three potential results: 1) intra-allelic recombination and exchange, 2) inter-allelic recombination and exchange, and 3) cryptic recombination (a process in which physical recombination takes place without the exchange of flanking genetic markers). (Bernstein et al., 1988).

While physical recombination is often referred to merely as "recombination," some authors also refer to the independent assortment of non-homologous chromosomes during meiosis as recombination. Such independent assortment can, and usually does, lead to new combinations of pre-existing alleles, whether physical recombination takes place or not (Michod and Levin, 1988). Finally, outcrossing may also be referred to as recombination, and may also serve to increase the number of new combinations of alleles (Michod and Levin, 1988).

Given the wide range of possible meanings for terms such as "sex" and "recombination," it is easy to see how misunderstandings arise and how important it is for authors to be as specific as possible when using such terms.

We have chosen to subdivide the many theories about the evolution of sex and recombination into two main categories: 1) those dealing primarily with the origin of sex, and 2) those dealing with its maintenance once it has arisen. Maintenance categories have been subdivided further, depending on whether or not they involve the removal of linkage disequilibrium (a distinction due to Felsenstein (1988)). Those that do involve the removal of linkage disequilibrium have been subdivided yet further, following Felsenstein's reasoning, into Fisher–Muller type hypotheses and Sturtevant–Mather type hypotheses. Other classification schemes that have been proposed are discussed.

THE ORIGIN OF SEX-RELATED PHENOMENA

The Origin of Genetic Transformation

We begin with the simplest mechanism by which genetic information can be transferred from one cell to another, or from the external environment into a cell. As will be seen, this process, which is likely to be one of the very earliest in the evolution of sex-like phenomena, provides clues to the later elaboration of sex and sex-like mechanisms.

Transformation is the uptake by a cell of exogenous DNA present in the surrounding medium (Stewart and Carlson, 1986). Transformation can occur naturally, or be induced artificially through the use of chemical or physical treatments.

The process can be broken down into four basic steps: 1) the development of competence, 2) the binding of DNA, 3) the internalization of the bound DNA, and 4) the recombination of the internalized DNA into the cell's genome.

In order for a cell to be able to undergo transformation, it must first become "competent." Competence is the physiological state that enables the cell to bind and take up exogenous DNA in a manner that protects the DNA from exogenous DNAses (Stewart and Carlson, 1986; Venema, 1979). Competence is heritable (Stewart and Carlson, 1986), and all naturally transforming bacteria (with the exception of *Neisseria gonorrhoeae*, which is constitutively competent (Sparling, 1966)) regulate their competence (Goodgal, 1982; Stewart and Carlson, 1986).

Natural competence for transformation is widespread throughout the kingdom monera, and is known to occur in the following genera: *Achromobacter* (Juni and Heym, 1980), *Acinetobacter* (Juni and Janik, 1969), *Azobacter* (Page, 1981), *Bacillus* (Goldberg et al., 1966), *Campylobacter* (Wang and Taylor, 1990), *Chlorobium* (Ormerod, 1988), *Deinococcus* (*Micrococcus*) (Tirgari and Moseley, 1980), *Haemophilus* (Albritton et al.,

1984; Gromkova and Goodgal, 1979; Mathis and Scocca, 1982), *Halobacterium* (Mevarech and Werczberger, 1985), *Helicobacter* (Haas et al., 1993), *Lactobacillus* (Hofer, 1985), *Methylobacterium* (O'Conner et al., 1977), *Methanobacterium* (Worrell et al., 1988), *Methanococcus* (Bertani and Baresi, 1987), *Moraxella* (Juni, 1974; Juni, 1977; Juni et al., 1988) *Mycobacterium* (Norgard and Imaeda, 1978), *Neisseria* (Catlin, 1960; Mathis and Scocca, 1982), *Nostoc* (Trehan and Sinha, 1981), *Pseudomonas* (Carlson et al., 1983), *Rhizobium* (Courtois et al., 1988), *Staphylococcus* (Rudin et al., 1974), *Streptococcus* (Behnke, 1981; Morrison et al., 1982; Shah and Caufield, 1993), *Streptomyces* (Roelants et al., 1976), *Synechococcus* (Chauvat et al., 1983; Shestakov and Khyen, 1970; Stevens and Porter, 1986), *Synechocystis* (Grigorieva and Shestakov, 1982; Lorenz and Wackernagel, 1993), *Thermoactinomyces* (Hopwood and Wright, 1972), *Thermus* (Koyama et al., 1986), *Thiobacillus* (Yankofsky et al., 1983), and *Vibrio* (Frischer et al., 1990).

In *Acinetobacter*, *Anacystis*, *Azotobacter*, *Staphylococcus*, and *Streptococcus* competence begins to develop with cell growth, and reaches its maximum during early to late log phase, while in *Bacillus*, *Chlorobium*, *Methylobacterium*, *Pseudomonas*, *Synechocystis*, and *Vibrio*, competence typically does not reach its maximum until the transition from log to stationary phase (Lorenz and Wackernagel, 1994). In other genera, such as *Agmenellum*, *Deinococcus*, and *Mycobacterium*, competence occurs throughout the log phase, declining in stationary phase (Lorenz and Wackernagel, 1994).

In some gram-positive bacteria, such as *Streptococcus* (Pakula and Walczak, 1963), *Pneumococcus* (Tomasz, 1965; Tomasz, 1966; Tomasz and Hotchkiss, 1964) and *Bacillus* (Akrigg and Ayad, 1970; Akrigg et al., 1967; Felkner and Wyss, 1964; Streips and Young, 1971), competence is induced by a small polypeptide excreted during the growth of a culture. These factors build up in the culture, until they reach a concentration sufficient to induce competence (Stewart and Carlson, 1986).

In some gram-negative bacteria, such as *H. influenza*, competence is internally regulated in response to a variety of environmental cues, all of which are commonly associated with the transition from exponential growth to stationary phase (Stewart and Carlson, 1986). Surprisingly, the presence of free DNA in the medium does not induce competence in any of the cases in which competence has been thoroughly studied (Stewart and Carlson, 1986).

The development of competence and the ability to be transformed is neither a simple nor a passive phenomenon. While a single competence factor has been identified in *Streptococcus*, as many as fifteen other proteins may be associated with the development of competence (Morrison et al., 1982). Further, at least in *B. subilis*, the transport of DNA across the cytoplasmic membrane is energetically expensive, and involves transport systems (Van Nieuwenhoven et al., 1982).

Once competence has developed, binding of the exogenous double-stranded DNA at the cell surface must occur before it can be taken up (Stewart and Carlson, 1986). In both gram-positive and gram-negative bacteria, once binding has been achieved, DNA enters the cell in a single-stranded form which is protected from digestion by restriction enzymes by one of several means (Stewart and Carlson, 1986). Once internalized, the transforming DNA can become integrated into the genome through single-strand displacement of the native homologous strand. The displaced strand is then degraded, and its constituents presumably are reused. In *B. subtilis*, around 70% of homologous DNA that is taken into the cell becomes integrated, and the average length of the integrated segments is about 8.5 k.b. (Dubnau, 1991).

A number of hypotheses have been put forward to explain why natural transformation might have evolved in the first place (Bodmer, 1972; Hoelzer and Michod, 1991; Michod et al., 1988; Redfield, 1988; Redfield, 1993a; Redfield, 1993b; Wojciechowski et al., 1989). Possible reasons for the evolution of transformation include the production of genetic variation through gene transfer (in a manner analogous to sex) (Stewart and Carlson, 1986), the need for DNA components either as a food resource or for DNA synthesis (Bodmer, 1972; Redfield, 1993b), and as a mechanism for avoiding the deleterious effects of mutation (Redfield, 1988, 1997) or for the repair of double stranded breaks in DNA (Hoelzer and Michod, 1991; Michod et al., 1988; Wojciechowski et al., 1989).

The Gene Transfer Hypothesis for the Origin of Genetic Transformation

It has been suggested that natural transformation may have evolved in order to facilitate the transfer of genetic information between cells (Stewart and Carlson, 1986). The fact that a number of naturally transformable bacterial species exhibit preferential binding and uptake of homologous DNA suggests that transformation may play a role in intraspecific genetic exchange. *H. influenzae* preferentially binds DNA containing an 9 bp recognition sequence that is found in 1471 copies per genome, or an average of one copy every 1.2 k.b. (Smith et al., 1999a; Sisco and Smith, 1979; Stewart and Carlson, 1986). *Neisseria gonorrhoeae* preferentially binds sequences which contain a 10 bp recognition sequence (Goodman and Scocca, 1988). This same sequence occurs 1891 times in the *Neisseria meningitidis* genome which also happens to average of about one copy every 1.2 k.b. (Smith et al., 1999a) and there is evidence that other genera such as *Campylobacter*

(Wang and Taylor, 1990) and *Pseudomonas* (Carlson et al., 1983) also preferentially bind and take up homologous DNA. For a review on DNA uptake signals in naturally transformable bacteria see Smith et al. (1999b).

It is noteworthy that some of the same bacteria that undergo transformation in nature also excrete DNA into their surroundings at the same time that they become competent for transformation (Sinha and Iyer, 1971). Plasmid-bearing strains have been shown to excrete plasmid DNA in addition to pieces of their chromosomal DNA (Lorenz et al., 1991).

Sinha and Iyer (1971) found that in *B. subtilis* this DNA excretion was also accompanied by the excretion of RNA and protein. They maintained, however, that the release of these macromolecules is a result of the development of competence, and not simply an artifact brought about by the "physiological stresses imposed by competency [regimes]". They supported this contention by pointing out that 1) similar results were obtained with two different strains and two different competency regimes, 2) a transformation-minus mutant which showed normal DNA uptake but was otherwise incapable of being transformed showed no sign of DNA release, and 3) a reversion of the transformation-minus mutant showed the normal phenotype with respect to release of DNA upon attainment of competence (Sinha and Iyer, 1971).

If autolysis of a portion of the cell population was responsible for the appearance of chromosomal as well as plasmid DNA in the growth medium, then one might expect to find intra-cellular enzymes in the medium as well. Lorenz et al. (1991) found that the levels of intra-cellular enzymes were low to undetectable during the time at which cells were releasing DNA into the surrounding medium. They also found that some *B. subtilis* strains released a "pulse" of extracellular DNase in the phase prior to the development of competence, and the release of DNA. They point out that since *B. subtilis* is not specific in its DNA uptake, this pulse of DNase could act to clear the environment of heterologous DNA, thereby paving the way for a "highly coordinated transformational gene transfer . . ." (Lorenz et al., 1991). Stewart and Carlson (1986) had previously pointed out that genetic exchange could easily be promoted through such a self-regulated release of genetic information, timed to occur at exactly the time at which neighboring cells become competent to take up such DNA.

It is likely that *B. subtilis* is not the only bacterium to excrete DNA into its surroundings. Extracellular DNA has been isolated from species in the following genera: *Alcaligenes*, *Brucella*, *Flavobacterium* (Catlin, 1956), *Micrococcus* (Takahashi and Gibbons, 1957), *Pseudomonas* (Catlin, 1956), *Streptococcus* (Ottolenghi and Hotchkiss, 1960). *Agrobacterium tumerfaciens* is also know to excrete DNA into plant cells, thereby inducing crown gall formation (Anker et al., 1971; Stroun et al., 1970), and both *E. coli* and

Pseudomonas fluorescens have also been shown to release DNA into plant cells (Stroun et al., 1970).

Having discussed some of the circumstantial evidence in support of transformation as a mechanism of genetic transfer, is there any empirical evidence that such exchange is important, or that it occurs in nature? The analysis of a 28k.b region containing *tbpAB* (transferrin binding protein) operon, *opaA* (an adhesin) and several housekeeping genes from serogroup A, subgroup IV-1 of *Neisseria meningitidis* revealed widespread exchange of chromosomal fragments among *N. meningitidis* as well as other *Neisseria* species. Of 98 strains examined, 19 had unique imported DNA fragments ranging in size from 1.5 to 9.9k.b., with a median size of 5.1k.b. In addition, all but one of the imported fragments contained the entire coding region of the *tbpB* gene, and the remaining fragment contained most of the gene. It was found that not only did *N. meningitidis* incorporate large fragments from other conspecific strains, but it also did so from other species of *Neisseria*. This process probably occurred through transformation during cohabitation of the nasopharynx, and the very high frequency of incorporation of fragments carrying a *tbpB* allele was most likely due to immune selection (Linz et al., 2000).

Another naturally transformable bacterial species which shows evidence of transformation mediated lateral gene transfer is *Helicobacter pylori*. *H. pylori* is a common gram-negative human pathogen belonging to the phylum proteobacteria. As much as 50% of the human population is believed to be infected with this bacterium, and of these individuals, approximately 10% are expected to develop peptic ulcers as a result of their infection (Suerbaum and Achtman, 1999). In *Helicobacter*, the genes *flaA* and *flaB* code for the two principle components of the the flagella, and the *vacA* gene produces a secreted cytotoxin. A study of genetic polymorphism in strains isolated from patients in Africa, Germany and Canada revealed a remarkable level of variability in these three genes with strains differing on average by 15–20% at synonymous sites, but only by 0.3–2.5% at non-synonymous sites (Suerbaum et al., 1998). Maynard Smith and Smith, (1998) designed a homoplasy test to detect recombination in gene sequence data from closely related organisms. Results from this test can range from H = 0 for a completely clonal population with out any recombination to H = 1 for a population exhibiting free recombination and linkage equilibrium. Application of this test to the *Helicobacter* data showed that "horizontal genetic exchange in *H. pylori* is so frequent that different loci and polymorphisms within each locus are all at linkage equilibrium." (Suerbaum et al., 1998). What is truly remarkable is the fact that *Helicobacter* showed higher values of H than *Drosophila*, a species which recombines every generation. The authors of this study concluded that *H. pylori* is panmictic-a rather startling finding for what traditionally would be considered an "asexual" organism.

A second more extensive study, which included portions of seven housekeeping genes, along with two pathogenicity associated genes, *cagA* and *vacA* corroborated the majority of findings from the first study. Most of the genes examined exhibited free recombination, a pattern which was superimposed upon two weakly discernable clonal groupings—one having an asian origin. Interestingly, some alleles of *cagA* (a virulence associated gene located within a pathogenicity island) showed almost no signs of recombination, and in fact, had the lowest H value of any species thus far examined (Achtman et al., 1999). The reasons for this apparent lack of recombination of *cagA* amidst a sea of recombination is unclear, but may represent selection for particular *cagA* alleles in certain geographic regions.

At a more localized level, recombination has been detected between different clones within a single human host. Kersulyte et al. (1999) documented six separate recombinational exchanges between two coinfecting clones of *H. pylori* in a Lithuthian patient. They suggest that such genetic exchange may allow the production of strains that are better able to parasitize their host than either of the parental strains from which they were derived.

In summary, in at least some of the naturally transforming bacteria examined, transformation is directly responsible for the production of genetically diverse progeny. Whether the production of genetic variability is the sole reason for the evolution of transformation, or is only one of several reasons, is less clear.

The Double-Strand Break Hypothesis for the Origin of Transformation

The early earth's atmosphere provided little protection from UV radiation, which causes the formation of thymine dimers as well as cross-links in DNA molecules. Double-stranded breaks can result as a byproduct of the repair of such cross-links. Double-stranded breaks can also result if dimers form adjacent to each other, or in close proximity to each other on opposing DNA strands. It is double-stranded breaks from these as well as other sources such ionizing radiation, and endogenous oxidative-radicals that have been suggested by several authors to be the principal impetus for the evolution of transformation (Hoelzer and Michod, 1991; Michod et al., 1988; Wojciechowski et al., 1989).

In yeast, the numbers of closely opposed dimers increase *linearly* with increasing UV for doses between 0 and 200 J/m^2 (Reynolds, 1987). In UV irradiated cells, thymine dimers can be removed by photoreactivation mechanisms which involve the use of visible light. In the absence of visible light, the repair of thymine dimers is thought to proceed by one of two mechanisms. Bresler (1975) has proposed a mutation-prone model, in which repair of a DNA lesion by the *Rec* system results in a point mutation. This can occur because the *Rec* system of repair removes a large segment from the damaged strand (1,000–2,000 b.p.) and re-synthesizes it using the opposing strand as a template. If there is another lesion somewhere within the 1–2 k.b. segment of the template the *Rec* system will randomly substitute a nucleotide opposite that lesion, resulting in a point mutation in the first strand (Bresler, 1975).

The second repair route (which is more important for our discussion) is through the action of excision repair enzymes which simultaneously chop out the opposing single-stranded sections of DNA that contain dimers (Reynolds, 1987). (In *E. coli* the *UvrABC* endonuclease removes a 12 base segment surrounding the dimer (Watson et al., 1987).) The simultaneous excision of both of the lesions will result in the production of a double-stranded gap (Reynolds, 1987).

According to Bernstein et al. (1987), the only way to repair such gaps is through recombinational repair using a homologous DNA molecule as an information template. It has been proposed that the evolutionary function of transformation is to provide just such an information template for use in the recombinational repair of DNA damage (Hoelzer and Michod, 1991; Michod et al., 1988; Wojciechowski et al., 1989). These authors have shown that the rate of homologous transformation increases in *B. subtilis* with increasing UV damage prior to transformation (that is, the ratio of transformants to total cells increases) (Michod et al., 1988). The Michod group also demonstrated that there is no increase in transformation rate of UV damaged cells when non-homologous DNA is used in place of homologous DNA (Wojciechowski et al., 1989), and that the proportion of *B. subtilis* transformants continues to increase with increasing UV irradiation, even when the source of transforming DNA is itself UV-damaged (Hoelzer and Michod, 1991). The reason that recombining with damaged exogenous DNA might still be advantageous is that the likelihood that the incoming damaged DNA will be damaged in the same location as the "resident" DNA is very remote (Hoelzer and Michod, 1991).

Mongold (1992) attempted to extend the results of Michod et al. to *Haemophilus influenzae*, a naturally competent, gram-negative bacterium. She found, as had Michod et al., that following UV induced damage, the survival of partially competent cultures was significantly higher when transformed with homologous DNA than when not transformed ($p < 0.0001$). In addition, there was no increase in survival among cells transformed by non-homologous DNA. Surprisingly, treatments involving fully competent cultures showed no significant effect of transformation on cell survival. Mongold found that transformation with a 9k.b. fraction of the 1.8M.b. *H. influenzae* genome was as effective as transformation with fragments representative of the entire genome- a finding which is difficult to explain. However, since no attempt was made to control for the presence of exogenous *Haemophilus* DNA in the cultures being transformed, it is difficult to assess what effect the presence of such DNA may have had on the results obtained.

Durand et al. (1993) examined the effects of UV irradiation on the transformation rates, survival and mitotic recombination rates of isogenic sexual (*Mat a/α*) and asexual (*Mat a/a*) strains of the yeast *Saccharomyces*. In the absence of photoreactivation (the light mediated reversal of pyrimidine dimerization), *Mat a/α* strains showed a significantly higher survival rate than isogenic *Mat a/a*, a finding consistent with their greatly enhanced ability to recombine and repair DNA damage. When UV irradiated strains were kept in the dark following irradiation, thereby preventing photoreactivation, both the *Mat a/α* and the *Mat a/a* strain showed a highly significant increase in their mitotic recombination rates as determined by the rate of reversion to prototrophy of a heteroallelic *ura3* marker. This finding may be indicative of the recombinational repair of double-stranded breaks resulting from the attempted repair of UV induced damages. Most important was the finding that when cells were pre-exposed to 100 joules/m^2 of UV, the sexual *Mat a/α* strain showed a greatly enhanced ability to transform plasmid DNA, but only when subsequently kept in the dark. The asexual strain showed no such UV induced enhancement of transformation efficiency, even at doses of up to 300 J/m^2. This indicates that in the absence of photoreactivation, the sexual strain greatly increased its rate of transformation at a time at which there would be double-stranded breaks present in the genome. This is exactly what would be predicted by the DNA repair hypothesis, although as discussed below, these results are not inconsistent with the nutritional hypothesis also.

DNA damage may not always lead to increased rates of transformation. Redfield found that DNA damage caused by UV or by mitomycin C did not induce competence in either *H. influenzae*, or *B. subtilis*, a finding that is in conflict with the predictions of the DNA repair hypothesis, and the findings of both Michod et al. and Mongold (Redfield, 1993a; Redfield, 1993b).

Judging from the conflicting and unpredicted results of some of the aforementioned experiments, this area of research needs more, well designed experiments before DNA repair can either be accepted or rejected as the evolutionary impetus behind the evolution of genetic transformation.

The Mutational Load Hypothesis for the Origin of Transformation

Redfield (1988) suggested that prokaryotic transformation may have evolved in order to reduce the mutational load. If a given microbe species is common in a particular environment, it is probable that exogenous DNA taken up by a cell will be from a dead conspecific. This will increase the likelihood that the DNA taken up will have regions homologous to the mutated section. It is also reasonable to assume that exogenous DNA has been exposed to similar if not greater levels of mutation, leading one to ask whether such DNA would be of any use. Redfield (1988) has developed a model in which she examined cases in which the source of transforming DNA was from cells killed by mutations. If the transformation is 'regulated' (that is, when the most fit cells are not transformed), then transformation always results in populations that are more fit than non-transformed populations at equilibrium.

A more recent analysis of a more biologically plausible situation produced rather different conclusions. Redfield et al. (1997) modeled a continuous time, mixed population, with overlapping generations, having both transforming individuals as well as non-transforming individuals. Both classes of individual were subject to deleterious mutation. Transformers could take up and integrate mutated DNA as well as wildtype DNA from non-transformers. The latter DNA would, with a certain probability, contain a null allele for competence; when integrated, such DNA would cause the transformer to lose its ability to transform. It was found that there were no plausible conditions under which an allele conferring competence could invade a non-transforming population. Redfield et al. concluded "... it seems likely that transformation has evolved for reasons other than reducing mutational load."

The Nutritional Hypothesis for the Origin of Transformation

Bodmer (1972) was the first to suggest that competence may be a means of sequestering precursors for DNA synthesis. More recently, Redfield (1993b) has been the chief proponent of a nutritional explanation for the evolution of transformation, pointing out that "the sugars, phosphates, and nitrogenous bases of DNA are nutrients in great demand." As was noted earlier, many naturally transformable bacteria grown in liquid media, become competent (capable of taking up exogenous DNA) at or near plateau (a time at which nutrients presumably become limiting) (Redfield, 1993b). In *B. subtilis*, several of the genes responsible for determining competence are also involved in nutrient uptake or processing (Dubnau, 1991). Competence in *Haemophilus influenzae* is induced by transfer to a starvation medium lacking both a carbon source and nucleotide precursors.

When DNA is taken up by starved cells, one strand is always quickly degraded, and the resulting nucleotides are released to the surrounding medium (Dubnau, 1991). If the DNA is not from the same, or a closely related species, then the remaining strand is also degraded and its constituents are reused (Pifer and Smith, 1985; Redfield, 1993b). If, on the other hand, the DNA is from a closely related species, then it may recombine with a homologous stretch of DNA in the cell's genome. If this occurs, Redfield points out, this does not prevent the cell from profiting from the nutrient content of the transformed DNA. The reason is that, while one strand of the entering fragment may be incorporated into the genome of the transformed cell, the "resident" DNA will have a strand of equal length displaced by homologous recombination, and it is this strand that will be degraded and its components reused (Redfield, 1993b).

Rather than an all or nothing approach, Redfield (1993b) favors a sequence of events by which transformation first appeared as a means of procuring valuable resources for both nutrition and DNA synthesis. As a fortuitous bonus, transforming cells would occasionally benefit from recombinational repair, and /or production of genetic diversity.

Dubnau (1999) points out several facts that he feels argue against the nutritional hypothesis for the evolution of transformation. The first of these is the fact that one of the two strands of DNA is always degraded and the resulting nucleotides are released to the external environment. This suggests that if transformation is an adaptation for the procurement of nucleotide resourses, it is not a very inefficient one. Secondly, many naturally competent bacteria only transform DNA containing conspecific recognition sequences. This sequence specificity would certainly not be expected under a purely nutritional model. Thirdly, Dubnau points out that *B. subtilis* is able to secrete powerful nonspecific nucleases into the environment, and has well developed systems for uptaking the resulting nucleolytic products. With such systems already in place, Dubnau asks the obvious question: "Why would the elaborate transformation machinery evolve to meet this need, which is met by a simpler and more generally useful pathwway?"

With these reservations aside, the nutritional model is nevertheless an intriguing hypothesis. The yeast *S. cerevisiae* may shed more light on this process. As previously discussed, in the yeast *S. cerevisiae*, Durand et al. (1993) found evidence that may lend support to both the DNA repair hypothesis as well as the nutritional hypothesis for the evolution of transformation. They found that in the absence of visible light, there is UV-enhanced transformation of exogenous DNA, a finding which directly supports the DNA repair hypothesis. To see why this finding may also lend support to the nutrition hypothesis we first need to ask why might it be advantageous for yeast with UV-damaged genomes to transform exogenous DNA, but *only* when visible light is not present? To suggest an answer to this question, it may help to examine some additional facts. As we have already discussed, UV irradiation produces cross-links and pyrimidine dimers, the latter of which are primarily responsible for the killing observed in UV-irradiated yeast (Kunz and Haynes, 1981a). One of the principle types of dimer so formed is the thymine dimer (Watson et al., 1987), in which two tandem thymine nucleotides are covalently linked together by a high energy bond. Such bonds are energetically costly to break in the absence of visible light, since under such conditions, photo-reactivation cannot occur. In the dark, thymine dimers are excised, leaving single stranded gaps. Double stranded gaps can form if the dimers are nearby on adjacent strands, and these gaps are enlarged by subsequent degradation (Reynolds, 1987)).

What happens to excised dimers in the absence of visible light? Can they be recycled? Or are they essentially unusable? Does this leave the cell "starved" for thymine? As strange as this sounds, this may be plausible. In *Saccharomyces*, the *only* way to synthesize thymidylate (dTMP) is via the methylation of deoxyuridine-5′-monophosphate (dUMP) (Kunz and Haynes, 1981a). This reaction is energetically expensive (Stryer, 1988). Kunz and Haynes (1981b) state that "any agent which tends to block DNA replication can be recombinagenic." Starvation for thymine nucleotides can block DNA replication, and has been shown to induce large increases in mitotic recombination in *Saccharomyces* (up to several orders of magnitude over background levels) (Kunz et al., 1980; Kunz and Haynes, 1981a). These authors suggested that this increase may be due to the misincorporation of uracil in place of thymine, with the subsequent removal of the uracil by uracil-DNA glycosylase. If endogenous dTTP pools become depleted, uracil excision repair will be incomplete, resulting in numerous single-stranded gaps which may act as substrates for recombination.

It is of interest to note that in the experiments of Durand et al. (1993), mitotic recombination was higher in the dark for both the sexual and asexual strains than in the light, a finding which may be indicative of starvation for thymine. Alternatively, as previously mentioned, this finding could be indicative of the production and repair of double-stranded breaks induced by UV cross-links. In either case, in the presence of light, photoreactivation reverses thymine dimerization, thereby eliminating the production of double-stranded breaks, while simultaneously preventing the sequestration of thymine in the form of unusable dimers. Under conditions of thymine starvation, it would be advantageous to take up exogenous

DNA as a nucleotide "resource", especially when the nucleotide required is energetically costly to synthesize.

Eukaryotic and prokaryotic transformation are very different phenomena, and the transformation protocol used by Durand et al. (the lithium acetate method (Ito et al., 1983)) is not a condition that yeast are likely to meet in nature. Therefore, we may only have been looking at an artifact of the transformation protocol. However, the fact that transformation can be induced in *Saccharomyces* by UV irradiation (a phenomenon also seen in *B. subtilis* (Hoelzer and Michod, 1991; Michod et al., 1988; Wojciechowski et al., 1989) and *H. influenzae* (Mongold, 1992)), suggests that the phenomenon may be more than just an experimental artifact. Interestingly, it has recently been found that yeast are transformable under more natural conditions, involving high concentrations of sugar in the absence of any other form of nutrient (Nevoight et al., 2000). As almost nothing is known about *Saccharomyces* in the wild, it will be important to determine how frequently transformation occurs in nature, and whether it can be induced by UV irradiation, starvation for nucleotides, or any other DNA-damaging or replication-inhibiting factor.

It should be noted at this point that while the observations of Hoelzer and Michod, 1991; Michod et al., 1988; Wojciechowski et al., 1989, and Mongold, 1992, do lend support to the DNA repair hypothesis, these same observations do not necessarily contradict the nutritional hypothesis.

THE ORIGINS OF SEXUAL RECOMBINATION

DNA Repair Hypotheses

Although the DNA repair hypothesis appears under our Origin of Sex heading, it certainly is also a Maintenance of Sex hypothesis and will appear in that later section.

While Dougherty (1955) was one of the first to propose DNA repair as an impetus for the evolution of sexual recombination, others have proposed similar concepts (Bengtsson, 1992; Walker, 1978). Nevertheless, it is Bernstein et al. who have most fully developed this concept, and it is their version which will be discussed here in detail. (Bernstein, 1987; Bernstein and Bernstein, 1991; Bernstein and Chen, 1987; Bernstein and Johns, 1989; Bernstein et al., 1985a; Bernstein et al., 1984; Bernstein et al., 1985b; Bernstein et al., 1985c; Bernstein et al., 1985d; Bernstein et al., 1987; Bernstein et al., 1988; Bernstein et al., 1981; Hopf et al., 1988; Michod, 1993; Michod, 1990; Michod and Gayley, 1992; Michod et al., 1988).

As already mentioned, Bernstein et al. (1984) broadly define sex as "a process in which the genomes of two parents are brought together in a common cytoplasm to produce progeny which may then contain reassorted portions of the parental genomes." This definition can be applied to all forms of sex-like phenomena, including any primitive sexual cycles possessed by organisms of the RNA world. In a more restrictive sense, Bernstein et al. (1988) later define sex as being composed of "two basic processes: 1) recombination (i.e. physical recombination) and 2) outcrossing, in the sense that the two DNA molecules involved in recombination come from different individuals."

Evolution of Prebiotic Replicators

It has been shown experimentally that organic compounds can be synthesized from the gases present in the Earth's prebiotic atmosphere by the input of energy from virtually any source (e.g. lightning, volcanic activity, shockwaves, or UV radiation) (Eigen et al., 1981). It is also generally accepted that the key building blocks necessary for the evolution of life were present in the primitive "soup" from which life is thought to have arisen (Miller and Orgel, 1974). Eigen et al. (1981) proposed that the first genetic carriers of information were single-stranded RNA molecules (ssRNA), composed of a chain of nucleotides. Each nucleotide contains one of the four bases: adenine (A), guanine (G), uracil (U), or cytosine (C). The bases A and U can form complementary pairs by means of hydrogen bonding, as can the bases C and G. While double-stranded RNA and DNA do not form secondary three dimensional structures by folding back upon themselves, ssRNA does so readily by the formation of hydrogen bonds between the complementary bases (Eigen et al., 1981).

According to Bernstein et al. (1985b; Bernstein et al., 1987), these secondary structures formed by ssRNA were the first molecular phenotypes, and were determined by the genotype of the RNA strand. Different genotypes produced different folding configurations, which in turn had different degrees of stability. Some configurations were more resistant to hydrolysis than others (Eigen et al., 1981). These authors point out that it has been shown experimentally that even without enzymatic catalysis single-strand RNA self-replication can occur through complementary base pairing.

It seems reasonable to assume that not all sequences would be equally adept at forming stable configurations and at self-replication. G-C bonds are about 10 times as strong as A-U bonds. This allows greater replicative fidelity because strands rich in G and C bind their complementary bases much more strongly than do strands rich in A and U. Sequences with greater stability and replicative fidelity would produce more replicas of themselves than less fit sequences. Eventually, competition for resources such as activated RNA monomers (ATP, GTP, UTP, CTP) would lead to the formation of what Eigen et al. (1981) referred to as a quasispecies: a master sequence (i.e. the most fit sequence) along with its retinue of attendant mutants which are derived from it. The actual master sequence is rarely seen in the quasispecies distribution. It is well defined statistically however, being the sequence in which "each position is occupied by the nucleotide most frequently found at that position". It is from these quasispecies distributions that selection "chooses" the most fit replicators.

Evolution of Hypercycles

In primitive single-stranded RNA, the error frequency is proportional to the sequence length. Using this fact, Eigen et al. (1981) calculated that the maximum length attainable by a sequence rich in G and C would probably have been somewhere between 50 and 100 nucleotides in length. They went on to state that "while this is long enough to provide self stabilization through internal folding, it is probably not long enough to encode an efficient catalytic protein." They concluded after extensive investigation that the only way to overcome this error threshold was for quasispecies to form interdependent cooperative couplings, which they referred to as hypercycles.

Hypercycles are self-regulating and self-perpetuating. Eigen et al. gave the following example of a simple two component hypercycle: RNA sequence 1 encodes a protein capable of catalyzing the self-replication of RNA sequence 2, which in turn encodes a second protein capable of catalyzing the self-replication of RNA sequence 1. Thus, although normally these two sequences (actually quasispecies distributions) would be in competition with each other for resources, once hooked into a hypercycle they cooperate and simultaneously impose strict controls on the rate at which the other replicates. This enforced cooperation prevents one quasispecies from outcompeting the other, and thus hypercyclic cooperation greatly increased the amount of genetic information available to "code for a minimal starting amount of enzymatic function" by allowing the coexistence of different genetic sequences.

Encapsulation of Hypercycles

Once a hypercycle had become established, it would ostensibly be susceptible to parasitism by mutant "cheater" quasispecies which would be able to take advantage of the proteins produced by it (utilizing them for catalysis of self-replication) without producing anything in return

(Bernstein et al., 1984). These mutant sequences would be able to replicate more rapidly than the hypercycles they were parasitizing, and eventually would drive both themselves and their host hypercycles to extinction. In order to circumvent this obstacle, hypercycles would have to become encapsulated, thereby confining the proteins they produce and preventing their use by non-cooperative quasispecies. Encapsulation would also greatly improve self-replicating efficiency by keeping the proteins that were produced confined, so that the presence of the required proteins would be assured rather than left to chance.

Encapsulation would require a fairly complex hypercycle; one which could encode for encapsulation, self-replication, and subsequent binary fission into "progeny" capsules. How so complex a hypercycle could come about is moot. However, Bernstein et al. (1984) carried out a computer simulation of competition for resources within a population of single-gene hypercycles, each of which produced a protein having random effects on other hypercycles. They found that approximately one percent of the time a composite hypercycle of four of more genes was produced. Thus, given enough time, it may have been possible for a hypercycle to develop a sufficiently complex protein-encoding ability to allow encapsulation to occur. Such a development is critical to the DNA repair hypothesis.

Effects of Genetic Damage

While encapsulation of a hypercycle into a primitive protocell allowed efficacious use of translation products, the protocell also had to survive the cost of genetic damage from sources such as UV irradation. The earliest atmosphere of the Earth was reducing, and could not provide the protection from UV radiation that our present atmosphere does.

It has been estimated that before the formation of a protective ozone layer, organisms exposed to sunlight would receive a lethal dose of UV radiation in less than 0.3 seconds (Sagan, 1973). It seems reasonable to assume that the damaging effects of radiation, along with various sources of chemical mutagenesis, would impose strong selective pressures upon these early protocells. The degree of selection depends upon whether one defines a hypercycle as being haploid or polyploid. According to Eigen et al. (1981), the first hypercycles must have been polyploid, composed of "hypercyclically organized quasispecies distributions." In such hypercycles, damaged sequences would simply disappear, and have their places taken by undamaged homologous sequences. Genetic damage would not pose too serious a problem for hypercycles organized according to the Eigen model, due to the high degree of redundancy provided by each quasispecies component. Bernstein et al. (1987), on the other hand, appear to suggest that the first encapsulated hypercycles were haploid when they state: ". . . encapsulation thus promotes efficient use of gene products, but at the cost of making the entire set of replicator segments vulnerable when only one is damaged or lost." Such would not be the case with Eigen's model.

Bernstein et al. go on to state: "This vulnerability could be reduced by maintaining more than one copy of each RNA segment in each protocell." This is explicit in Eigen's model. They then argue that such redundancy could be costly if it involved longer generation times and greater resource requirements. They propose that a major problem faced by early protocells might have been how to deal with genetic damage while keeping the cost of redundancy at a minimum. After performing a cost/benefit analysis, they conclude that the best possible strategy given this situation would be for each protocell to be haploid most of the time. Occasionally, they would undergo a simple sexual cycle in which two protocells fused to form a transient diploid allowing "mutual reactivation of lethally damaged protocells." The fact that some extant single-stranded RNA viruses, such as the influenza virus, use a form of reactivation similar to this primitive sexual cycle envisioned by Bernstein et al. (1985a; 1987) lends credence to this view.

The influenza virus is composed of eight single-stranded segments. When two or more viruses infect a single cell their segments are injected into the cell, where they then replicate. The progeny viruses released from such a multiple infection have composite genomes made up of a mixture of segments from the parental genomes. It has been found that if the viruses are UV irradiated prior to infection (thereby damaging their RNA), multiply-infected cells produce more viable progeny than singly-infected cells (Bernstein et al., 1985b; Bernstein et al., 1987). This result is due to the genetic reassortment of undamaged parental segments, yielding viable progeny with complete, undamaged genomes. This sex-like process has been termed multiplicity reactivation, and occurs in many viruses (see Bernstein et al., 1984; Bernstein et al., 1985b; Bernstein et al., 1987 for reviews).

Bernstein et al. (Bernstein, 1987; Bernstein and Chen, 1987) performed an interesting experiment with phage T4. Ordinarily, when a T4 phage infects *E. coli*, it produces an "immunity" factor which prevents superinfection of the same cell by another phage. Presumably this has been selected for because it allows the phage to monopolize the host's resources for its own asexual reproduction, so that it will not be required to compete with another phage for access to resources.

If, however, the infecting phage's genome is damaged in some way (such as by UV irradiation), then not only is the phage's viability decreased, but the expression of the element conferring immunity is also inhibited. This would allow superinfection of the cell by a second phage resulting in the sex-like process of multiplicity reactivation. It was found that in multiply infected cells, marker genes of the damaged phage were "rescued" more frequently than in cells which were singly infected. Thus, when T4 phages suffer genetic damage, they are more likely to reproduce sexually and thereby rescue themselves from certain genetic death. While these results conform to predictions of the DNA repair hypothesis, the possibility cannot be excluded that this sex-like reproduction is merely a result of damage to the gene or genes encoding the immunity function rather than a general response to genetic damage. Nevertheless, Bernstein and Chen (1987) did find that at high levels of UV damage, a marker gene may be "rescued" up to 6 times more efficiently by this sex-like process than is the case when the immunity function is simply disabled. This indicates that there is more to this phenomenon than simply the inactivation of the immunity function due to genetic damage.

To summarize, according to the repair hypothesis, once encapsulation had evolved there should be selective pressure to reduce redundancy. At the same time, there should be an increase in the intensity of selection for mechanisms to repair or remove damage. Taken together, these two opposing forces may have led to a "compromise" in the evolution of the first sexual cycle. Presumably, this compromise would entail a limited degree of redundancy during some part of this primitive sexual cycle. Such redundancy is necessary to repair double-stranded damage. This redundancy would be limited however, so as not to impair the protocell's fitness by making its replication too slow or costly a process. This primitive sexual cycle served as a means by which protocells would regain damaged genetic information, using a process which was probably similar in many ways to present-day multiplicity reactivation.

The Evolution of Genetic Segregation

Another problem faced by early protocells was the need to segregate the replicated genetic material into viable progeny. Assuming Eigen et al. (1981) were correct, and the first protocells consisted of a group of highly polyploid, cooperative quasispecies distributions, then simple binary fission could, on average, result in the segregation of at least one of each type of sequence into each progeny. This would work well enough as long as the total number of copies within each quasispecies remained high. As Bernstein et al. (1985b; 1987) pointed out, however, such redundancy has associated costs such as increased generation time and greater resource requirements. If we assume, as Bernstein et al. did, that selection will favor a reduction in redundancy, then we must also assume that some more accurate form of segregation must have evolved concomitantly.

The scenario that Bernstein et al. (1984b) propose to solve this segregation problem proceeds as follows: (1) The original segmented single-stranded RNA protocells evolved into segmented double-stranded protocells (double-stranded RNA being more stable), which underwent the same type of sexual union and multiplicity reactivation as their predecessors. Because the genome was segmented, successful segregation was still left to chance. Contemporary examples of this stage of evolution are the reoviruses (Bernstein et al., 1984b; McClain and Spendlove, 1966). (2) In the next step, genes with related functions were grouped together on the same colinear RNA duplex. This partially solved the segregation problem, by making sure that genes critical for a particular function would segregate together. Bacteriophage ϕ6 resembles this stage, since it has three RNA duplex segments, each containing three to four genes (Cupples et al., 1980). (3) Double-stranded RNA was subsequently supplanted by DNA, presumably because the latter is more stable. (4) Finally, in order to ensure greater accuracy of segregation, genes became covalently linked in a linear sequence. This stage was accompanied by the evolution of enzymes catalyzing recombination, which allowed sequences to be shuffled, and damaged regions to be removed. Bacteriophage T4 resembles this stage of evolution. T4 has a continuous DNA genome, carrying genes for enzymes involved in recombination that can mediate multiplicity reactivation between damaged bacteriophages.

To recapitulate, the evolution of segregation involved several concurrent trends: a shift from less stable to more stable encoding sequences (i.e. RNA → DNA), a shift from highly segmented to essentially continuous genomes, the evolution of recombination enzymes with which to repair these continuous genomes, and finally, a transition from stochastic segregation to the more precise deterministic segregation.

The Importance of Genetic Alterations

Two types of genetic alteration are pertinent to the repair hypothesis, genetic damage and mutation. As defined by Bernstein et al. (1985c), genetic damage can be defined as "physical alterations in the structural regularity of DNA such as breaks, depurinations, depyrimidinations, crosslinks, thymine dimers, and modified bases. Mutations are defined as "changes in the base-pair sequence of DNA that result from substitution, addition, deletion, or rearrangement of the standard base pairs." Genetic damage is easily recognized by repair enzymes because it involves fundamental changes in DNA structure. Mutation, on the other hand, involves no such changes in structure, simply being a process by which information is randomized by addition, deletion or substitution of bits of genetic code. As such, there is no way for enzymes to distinguish a mutant sequence from a "correct" sequence, and the only way mutants can be recognized is through the process of natural selection. According to the repair hypothesis, recombination evolved as a way of removing genetic damage, while diploidy and outcrossing evolved in order to mask deleterious recessive mutations through genetic complementation.

Evolution of Recombinational Repair

Genetic damage can involve both strands of the DNA or be confined to single strands. Single-strand damage can be repaired by excision repair mechanisms, which use the undamaged complementary strand as a source of information from which the excised portion of the damaged strand can be reconstructed. It is double-strand damage however, which plays a central role in the repair hypothesis.

Double-strand damage can be caused by a variety of means. Ionizing radiation causes double-strand breaks, while UV radiation can induce DNA crosslinks (Bernstein et al., 1981). Two byproducts of cellular respiration are also known to cause double-stranded damage. These are the superoxide radical (O_2^-) and hydrogen peroxide (H_2O_2) (Bernstein et al., 1985c). It appears that naturally-occurring free radicals may cause a considerable amount of double-strand damage. A conservative estimate puts the number of double-stranded breaks per diploid human nucleus per day that are caused by H_2O_2 at about 0.1. This could impose considerable selective pressure, since as few as one unrepaired double-strand break can lead to the death of the cell. It must be noted, however, that the degree of lethality caused by double-strand breaks varies considerably from organism to organism. In *E. coli*, one double-strand break is lethal, while in the bacterium *Deinococcus radiodurans*, a LD_{90} dose of radiation produces about 1400 double-strand breaks (Resnick, 1976). During growth conditions *Deinococcus* is polyploid containing from four to ten copies of its genome per cell (Harsojo et al., 1981; Hansen, 1978). It is estimated that as many as 700 crossovers may form within the nucleoid during the repair of the DNA damage induced by exposure to 1.75 Mrads of gamma irradiation (Daly and Minton, 1995). It has been suggested that *Deinococcus* may has evolved such a formitable DNA repair system as an adaptation to desication which can induce double stranded breaks (Mattimore and Battista, 1996). As a comparison, the average diploid mammalian cell requires 50 to 200 double-strand breaks to induce death (Resnick, 1976). These differential susceptibilities to double-strand breaks clearly reflect different repair capabilities and therefore should not be taken as evidence that double-strand damage does not impose selective pressures on some organisms.

Unlike single-strand damage, double-strand damage cannot be repaired by excision repair. Since both strands are damaged, there is no template from which to copy. The only known means of repairing double stranded damage is through recombinational repair, which requires a second homologous DNA sequence that can serve as a template to provide undamaged information (Bernstein et al., 1987).

Bernstein and Johns (1989) performed an experiment to determine whether DNA damage could induce sexuality in the facultatively sexual yeast *Schizosaccharomyces pombe*. They found that treatment of late exponential-phase vegetative cells with H_2O_2 increased the proportion of cells undergoing mating and meiotic spore formation (i.e. sex) an average of eightfold over that found in cultures not exposed to H_2O_2. While this is exactly what would be predicted by the DNA repair hypothesis, it is impossible to exclude the possibility that such an increase in sexual reproduction is merely a generalized response to stressful conditions. Many organisms can be induced to undergo meiosis and spore formation in response to stressful conditions such as nitrogen starvation (Sager and Granick, 1954).

In response to this criticism, C. Bernstein has pointed out that such starvation causes DNA damage (pers. comm.). It has also been shown that double-stranded breaks can initiate recombination in *Saccharomyces* (Sun et al., 1989). This lends further credence to the view that recombination may have had its earliest origins as a repair mechanism capable of removing double-strand damage.

Thus, according to the repair hypothesis, by the time DNA had replaced its double-stranded RNA precursor, genomes were no longer segmented. Enzymes had to evolve which would allow these early cells to remove damaged DNA by two new means in addition to multiplicity reactivation: excision repair and, even more importantly, recombinational repair. According to Bernstein et al. (1981) the evolution of recombinational repair eventually led to the evolution of meiosis (although they did not speculate about the pathways by which this would have occurred).

The variation that is produced by genomic reassortment associated with multiplicity reactivation, and by the allelic exchange between homologous chromosomes that occurs occasionally during recombinational repair, is therefore viewed as being merely a byproduct of DNA repair mechanisms. It appears not to be the reason that recombination evolved (Bernstein et al., 1985c; 1987). This is because both the Holliday (1964) and the Szostak et al. (1983) models of meiotic recombination predict that 50% of recombination events should result in exchange of flanking markers, but selection seems in fact to have reduced this percentage. In ten different studies, encompassing five species of fungus, it was found that the average ratio of non-exchange to exchange of flanking markers in meiotic products was 66:34. (For a summary of these studies see Whitehouse (1982), tables 19 and 38). Kunz and Haynes (1981b) cite a similar figure for *Saccharomyces cerevisiae* in which they find a mean association of 32% for flanking marker exchange with gene conversion. Data from *Drosophila melanogaster* is very similar, with only about 33% of the meiotic recombination events at the *rosy* cistron showing exchange of flanking markers (Chovnick et al., 1970). A more recent study of the rate of flanking marker exchange associated with recombination in *Neurospora* suggests that for at least some loci, the rate of cryptic recombination may exceed 90% (Bowring and Catcheside, 1996).

A possible explanation for these figures is that selection is actually acting to reduce the amount of variation produced through recombination between flanking markers. This could be a means of reducing the recombinational load without losing any of the benefits of recombinational repair. (It should be reemphasized at this point that here we are discussing the recombination associated with physical crossing-over within chromosomes as opposed to the recombination arising from independent assortment of chromosomes during meiosis. The repair hypothesis has no bearing on the latter). In fact, Bowring and Catcheside (1996) went on to state "If there is a common intermediate for conversion events and crossover events, there must be a mechanism that biases choice in favor of preserving the parental configuration of flanking markers." According to the repair hypothesis, if recombination is an adaptation to produce genetic variation, one would not expect there to be a bias towards the preservation of parental linkage groups. Clearly, in a number of organisms there appears to be a bias towards the maintenance of parental gene configurations, a finding which lends credence to the view that physical recombination did not evolve as an adaptation to produce genetic variation.

In should be mentioned that when it comes to the role that sex plays in producing genetic variation, some feel that the most relevant type of sexual recombination is not physical recombination but rather that recombination resulting from the independent assortment of chromosomes (B. Charlesworth, AGA conference proceedings).

Evolution of Diploidy

A key assumption of the repair hypothesis is that as primitive haploid cells evolved, their environment became more and more complex. Adaptation to such an environment demands an increasing amount of information, and therefore more genes and a larger genome (Bernstein et al., 1981). As the number of genes increased from about 50 (in organisms having genomes commensurate in size with that of bacteriophage λ) to 50,000 (in organisms having genomes similar in size to that of *Neurospora crassa*), cells became more and more susceptible to genetic damage (Bernstein et al., 1985b). This resulted in selection for increasing accuracy of replication.

The observations of Drake (1974, 1999) are in accordance with the notion of selection favoring increasing accuracy of replication with increasing genome size. He found that, in six haploid organisms with genomes varying nearly ten thousand-fold in size (from 6.4×10^3 b.p. in phage M13 to 4.2×10^7 b.p. in *Neurospora crassa*), as the size of the genome increased, the error rate per base pair decreased. The total error rate per genome per generation was almost constant with a mean of 0.0034 (Drake, 1999).

According to the DNA repair hypothesis, however, this reduction in error rate could not have been limitless. A point was reached at which further reduction in the error rate would be selected against due to the increasing costs associated with it. These costs could be attributed to the increasingly complex enzymatic systems and/or the cost of slowing down the replication process as replication accuracy increases (Bernstein et al., 1981). This latter cost could be inferred from experiments with phage T4 showing that polymerase mutants with greatly reduced rates of mutation also had much slower replication rates (Gillin and Nossal, 1976). These results suggest that while lower mutation rates are possible, they are not selected for in natural populations.

The lower limit of mutation per base pair per replication shows less than a twofold difference between prokaryotes and eukaryotes, being approximately 5.4×10^{-10} in *E. coli*, and 3.4×10^{-10} in *Drosophila melanogaster* (Drake, 1999). In the long run, lineages with mutation rates approaching zero will not fare well in a changing environment, while those with higher (but not too high) rates of mutation are more likely to produce successful new genotypes (Bernstein et al., 1981).

There is some experimental support of the view that selection actually does, in some situations, favor higher rather than lower mutation rates. In chemostat competition experiments using *E. coli*, mutator populations have been shown to have higher fitness than wild type populations (Cox and Gibson, 1974; Gibson et al., 1970). This advantage was shown to be related to mutations which arose during the competitions, and was not simply the result of some inherent fitness advantage that had been possessed by the mutator strain at the outset of the competitions (Cox and Gibson, 1974). In the Lenski lab's long-term *E. coli* evolution experiments, following 10,000 generations of adaptation, one quarter of the twelve evolving lineages had mutator strains arise and sweep to fixation (Sniegoski et al., 1997). It is thought that the fixation of these mutator alleles may have been the result of genetic hitch-hiking of the mutators along with adaptive mutations. While this may be a potent force in the short term evolution of asexual populations, in sexual populations, recombination would break-up the association between a mutator allele and the beneficial mutant that it may have created. For this reason, hitch-hiking is unlikely to result in the fixation of mutators in sexual populations (Sniegowski et al., 2000). As Sniegowski et al. put it, "The physiological cost of reducing mutation below the low level observed in most populations may be the most important factor in setting the genomic mutation rate in sexual and asexual systems, regardless of the benefits of mutation in producing new adaptive variation."

According to the repair hypothesis, once this limit on the accuracy of replication is attained, selection nevertheless would continue to favor larger genomes in order to provide the necessary information for increasingly complex organisms. As the genome expanded, deleterious mutations would begin to accumulate because the number of mutations per genome per unit time is proportional to the size of the genome (Bernstein, et al., 1981; 1987).

This imposes a larger mutational load on organisms with larger genomes. For a haploid genome containing n loci with a mutation rate to slightly deleterious alleles of μ, and given the assumption of no linkage disequilibrium, then, at mutation-selection balance, the frequency of these deleterious alleles will be about μ/s where s is the degree to which each allele lowers individual fitness. If s is small, then the *average* fitness of the population (W) at equilibrium is approximately $e^{-\mu n}$ (Maynard Smith (1978)).

W decreases with increasing genome size. For this reason, Bernstein et al. (1985b; 1987; 1981) proposed that as genome size and the mutational load increased, selection began to favor masking of these deleterious recessive mutations via a transition to a life cycle in which the diploid phase predominated. Such a transition would, however, afford only transient relief from this mutational load. This is because, in a predominantly diploid life cycle, only homozygous recessives would be subject to elimination through selection. Thus, new mutations would accumulate due to the masking effect of genetic complementation in heterozygotes. Such mutations would continue to accumulate until they reached an equilibrium frequency of approximately $(\mu/s)^{1/2}$ (Bernstein, et al., 1981).

At equilibrium, the frequency of expression of these deleterious recessives is equal to $[(\mu/s)^{1/2}]^2$ or simply μ/s (Muller, 1932)—the same frequency of expression as in the ancestral haploid lineage lacking complementation. But there would be a period, perhaps quite lengthy, before this new equilibrium was reached. As envisioned by Bernstein, et al. (1981), the advantage associated with a shift to diploidy, even though it was only transient, could have led to a further enlargement of the genome. This would have occurred "primarily through a process of gene duplication and subsequent accumulation of mutations leading, ultimately, to divergence of function." Once diploidy evolves, it is very difficult for a lineage to return to a haploid state, because of the immediate costs associated with the haploid expression of the previously masked deleterious recessive alleles (Crow and Kimura, 1970).

Evolution of Outcrossing

As pointed out earlier, sexual reproduction can be considered to consist of two fundamental processes: (1) recombination and (2) outcrossing. In this section we turn to the possible influence of genetic complementation on the evolution of outcrossing.

By definition, haploid organisms must outcross if they are to take part in a sexual cycle. The matter is not so straightforward for diploid organisms. For recombinational repair to occur, homologous chromosomes must pair during meiosis. In diploid organisms, the two homologs could come from different individuals or from the same individual. Due to the fact that the major costs of sex are associated with outcrossing (the cost of mating, the cost of males, the cost of reduced relatedness between parent and offspring, and the recombinational load), one would expect most diploid organisms to employ some form of closed system in which both homologs are derived from the same individual (e.g. automixis or obligate selfing) (Bernstein et al., 1985b; Bernstein et al., 1981). Such is not the case. Since outcrossing systems are far more common than are closed systems, there must be advantages to outcrossing which outweigh these costs (Bernstein, et al., 1981; 1985b). According to Bernstein et al. (1987), these advantages stem from the fact that outcrossing reproductive systems combine the advantages of both recombinational repair and masking of deleterious recessives due to genetic complementation (Table 1 (taken from Bernstein et al., 1987)).

TABLE 1. Classification of Diploid Reproductive Systems

Reproductive system	Masking ability at equilibrium	Recombinational repair	Source of homologous chromosome
Automixis	Low ($\sim 2\mu$)	Yes	Self
Selfing	Low ($\sim 2\mu$)	Yes	Self
Outcrossing	Intermediate (1 to $(N\mu)^{1/2}$)	Yes	Another individual
Endomitosis	High ($\sim$N)	Limited	Self
Apomixis	High ($\sim$N)	No	N.A.
Vegetative	High ($\sim$N)	No	N.A.

(N denotes the number of functional genes per genome, which in higher organisms is approximately 40,000 or more).

According to this table, at equilibrium an outcrossing system not only provides recombinational repair but also provides a moderate degree of masking, being able to mask somewhere between 1 and $(N\mu)^{1/2}$ deleterious recessive mutations, where N is the number of functional genes per genome. If μ equals the rate of deleterious mutation per haploid genome per replication, then at equilibrium the mutational load of all the reproductive systems listed in table 1 is equal to $[1 - e^{-\mu}]$, and the survivorship of all these systems is simply $e^{-\mu}$ (Bernstein, et al., 1987). Although all the reproductive systems listed are competitively equal at equilibrium (the mutational loads are equal), there is a transient advantage to switching to a reproductive system with greater masking ability. Thus, there is a transient selective advantage for making the transition from selfing or automixis to outcrossing, so long as the benefits of recombinational repair are retained and so long as the short-term benefit derived from outcrossing outweighs the short-term cost of outcrossing. This benefit is analogous to the transient benefit of greater masking ability that is obtained from switching from a predominantly haploid life cycle to a predominantly diploid life cycle. It should be noted that table 1 ignores the effect of mitotic recombination, which is an important repair mechanism in diploids (see later discussion of the origin of mitotic recombination and repair).

Once a system changes to outcrossing (or to diploidy, for that matter), the short-term advantage associated with masking will gradually disappear as the number of deleterious recessive alleles increases to a new equilibrium value. At this point, the mutational load will be the same as before the transition, and the end result will be a net loss in individual fitness due to the additional costs associated with outcrossing. However, once outcrossing has evolved in a lineage, it will have great difficulty in making the transition back to its former reproductive system, because of the very high short-term cost associated with the unmasking of deleterious recessive alleles which have been hidden by the complementation afforded by outcrossing.

Mitotic Recombination and the DNA Repair Hypothesis

Meiotic recombination is not the only means of repairing double-stranded damage. Mitotic recombination can also repair double-stranded damage in diploid eukaryotic cells (Bell, 1988a; Bernstein and Bernstein, 1991; Bernstein and Chen, 1987). Homologous recombination events are commonly observed when linear DNA is introduced into animal cells in vitro (Wake et al., 1985), presumably resulting from the activity of systems for the recombinational repair of double-stranded breaks (Bernstein and Chen, 1987). Yeast mutants which were selected for their reduced levels of mitotic exchange were subsequently found to be UV and X-ray sensitive (Bernstein and Chen, 1987; Rodarte-Ramon, 1972; Rodarte-Ramon and Mortimer, 1972). *Rad9* mutations in *Saccharomyces* almost totally eliminate both spontaneous and UV-induced mitotic recombination. Such mutants also show greater sensitivity to UV irradiation (Kowalski and Laskowski, 1975). In a review of DNA repair in *Saccharomyces*, Friedberg (1988) cited two groups of *RAD* genes which are involved in recombinational repair. The first group, made up of *RAD51*, *RAD52*, and *RAD54*, is essential for both mitotic and meiotic recombination, as well as the repair of damage induced by ionizing radiation. The second group (*RAD50*, *RAD53*, *RAD55*, and *RAD56*) is also required. However, mutations in these genes do not have as severe an effect as do those in the first group.

It has also been shown that a mutant line of Chinese Hamster cells (EM9), which is significantly more sensitive to ionizing radiation than wild type (Thompson et al., 1982), also shows only about 38% of wild type levels of homologous recombination (Hoy et al., 1987; Thacker, 1989).

In the yeast *Saccharomyces*, mitotic recombination can be induced by a variety of means, including X-rays, UV, heat, gamma rays, heavy ions, nitrous acid, MMS, acridine orange, and starvation for various amino acids (Davies et al., 1975; Fabre and Roman, 1977; Fogel and Hurst, 1963; Henaut and Luzzati, 1972; Kunz and Haynes, 1981b). In general, any agent which tends to block DNA replication can be recombinagenic (Kunz and Haynes, 1981b). Fabre and Roman (1977) showed in an elegant experiment that radiation-induced recombinagenic lesions caused the production of a diffusable product which promoted recombination. They concluded that increased mitotic recombination rates reflect not a net increase in recombination rate in individual cells, but rather an increase in the proportion of cells capable of undergoing recombination. Induced rates of mitotic recombination can be up to 1000 times higher than spontaneous rates, and may in fact reach meiotic levels (Kunz and Haynes, 1981b). It has been found by other investigators that in *Saccharomyces*, *MAT* homozygotes (i.e. asexuals) show a markedly lower level of both spontaneous and induced mitotic recombination than *MAT* heterozygotes (i.e. sexuals) (Esposito and Wagstaff, 1981; Friis and Roman, 1968; Kunz and Haynes, 1981b; Lemontt, 1980).

We have conducted similar experiments (Birdsell and Wills, unpublished results) with four diploid strains constructed for competition experiments, two sexual and two asexual, all of which are heteroallelic for the *ura3* marker. We found that the spontaneous reversion rate to prototrophy of the sexual strains in treatments on four different media ranged from 1.32 to 3.58 times that of isogenic asexuals, with an average of 2.03 times that of the asexuals. The mean rate of reversion for the four heteroallelic strains was about 1 in 10^4 cells. This is the first time measurements have been made of the difference in mitotic recombination rate between truly isogenic sexual and asexual strains. It should be pointed out, however, that other loci may give different results, as there are "hot spots" of recombination throughout the genome. Further, *MAT* homozygotes (asexuals) are more sensitive to both ionizing radiation and MMS than heterozygotes (sexuals) (Kunz and Haynes, 1981a; Kunz and Haynes, 1981b; Lemontt, 1980). This can probably be attributed at least in part to their different rates of mitotic recombination.

Summary

According to the DNA repair hypothesis of Bernstein and collaborators, a simple sexual cycle evolved very early in the history of life as a means by which two segmented RNA protocells could repair genetic damage. The fusion of these two haploid protocells into a transient diploid allowed each to make use of the genetic redundancy provided by the other in order to regain lost genetic information. This occurred through a process similar to the multiplicative reactivation seen today in many viruses. Following mutual reactivation, this diploid form would have undergone binary fission to yield reactivated progeny. This stage posed a difficulty, due to the fact that segregation of all the necessary segments into each of the two progeny would be a matter of chance. This problem would have been gradually overcome by the evolution of a continuous DNA genome. This had its own problems, however, since the genome would have been subject to a variety of types of genetic damage which could no longer be removed without the aid of special enzymes.

The occurrence of double-stranded damage is particularly important to the repair hypothesis. The hypothesis suggests that recombinational repair evolved in order to make the repair of double-stranded damage possible. This eventually led to the evolution of meiosis. Bernstein and co-workers do not speculate on the exact mechanisms by which meiosis might have evolved from purely recombinational repair mechanisms.

While the repair hypothesis acknowledges that the variation sometimes generated by the recombinational repair process may be very adaptive in a variety of situations, its emphasis is on the evolution of repair mechanisms. Any benefits conferred by this variation are a byproduct of the repair mechanism, and are not the ultimate reasons for the evolution of recombination.

As primitive organisms became more complex, their genomes expanded. This expansion led to a concomitant increase in the number of mutations per genome. Any further increase in the accuracy of replication was too costly, and thus there was a gradual transition to predominantly diploid lifecycles. Diploidy afforded an advantage by masking deleterious recessive mutations through genetic complementation. This advantage was transient however. Mutations would accumulate until a new equilibrium was reached, at which point, the mutational load was equal to that of the ancestral haploid organisms. The transition to diploidy allowed a further expansion of the genome through gene duplication and subsequent divergence of function. Once a lineage had made the transition to diploidy, it would be virtually impossible to make the reverse transition because so many deleterious alleles would be uncovered.

The hypothesis supposes that a similar process was responsible for the evolution and maintenance of outcrossing. Haploid organisms must outcross if they are to employ a sexual cycle. Diploid organisms employing automixis or selfing would reap a short-term benefit by switching to outcrossing. This is because of the greater probability of complementation afforded by the fusion of two unrelated haploid genomes (i.e. there is a greater probability that the two genomes will not share the exact same mutations and hence each one will be able to make up for what the other lacks). Once a lineage is outcrossing, a switch to an asexual form of reproduction imposes stiff penalties in the form of the expression of deleterious recessive alleles. Such transitions are unlikely to occur unless the costs associated with outcrossing are much greater than the cost of making the transition (e.g. if mates are extremely rare).

The Margulis and Sagan Hypothesis for the Origin of Sex

Of all the theories discussed in this review, only three deal explicitly with the origin of sex: 1) the DNA repair hypothesis (Bernstein, 1987; Bernstein and Bernstein, 1991; Bernstein and Chen, 1987; Bernstein and Johns, 1989; Bernstein et al., 1985a; Bernstein et al., 1984; Bernstein et al., 1985b; Bernstein et al., 1985c; Bernstein et al., 1987; Bernstein et al., 1988; Bernstein et al., 1981), 2) the Contagion hypothesis (Hickey and Rose, 1988; Hickey, 1982; Hickey, 1993; Rose, 1983), and 3) the Origin hypothesis of Margulis and Sagan (1984; 1985). Of these, the theory of Margulis and Sagan provides the most detail of the physiological and morphological origins of cellular structures involved in sexual reproduction, while at the same time ignoring or reinterpreting conventional genetic and evolutionary mechanisms.

Margulis and Sagan define sex as "any process that recombines genes (DNA) in an individual cell or organism from more than a single source; sex may occur at the nucleic acid, cytoplasmic, and other levels." They make an important distinction between prokaryotic sex (conjugation, transformation, and transduction), and eukaryotic (meiotic) sex, and suggest that: "the origin of sex, *per se*, is a misnomer.... Many things fall into place if we separate the various elements that have formerly been joined under the single heading of sex. If, for instance, we accept that meiosis and syngamy are of separate origin from crossing-over, it immediately becomes apparent why crossing-over does not always accompany meiotic prophase." (Margulis and Sagan, 1984) p. 18.

They note the physical recombination of DNA, involving the breakage and union of DNA molecules, may be an evolutionary response to UV-induced damage in bacteria. In contrast, meiotic reduction division and fusion involve the union of entire intact DNA molecules within a common nucleus, and do not necessarily require the breakage of DNA.

Evolution of Bacterial Sex

It is thought that bacterial sex first evolved in the Archaean aeon between 3.5 and 2.5 billion years ago. Meiotic sex, on the other hand, probably did not first appear until the Proterozoic, between 2.5 billion and 580 million years b.p. All prokaryotes differ fundamentally from eukaryotes in that they lack a membrane-bound nucleus, and they do not have the true chromosomes (histone-DNA complexes) associated with eukaryotes. Instead, the bacterial genophore consists of naked DNA attached to membranes, and because it is much smaller in diameter than the eukaryotic chromosome, it is referred to as a chromonema ("colored thread"), as opposed to a chromosome ("colored body"). Prokaryotes reproduce through binary fission, while eukaryotic cells reproduce through the far more organized means of mitosis.

In some ways, Margulis and Sagan's hypothesis overlaps with the other two origin hypotheses. Margulis and Sagan credit DNA damage as being the primary impetus for the evolution of bacterial repair mechanisms, which in turn pre-adapted these prokaryotes to sexuality.

The strong selective pressures imposed on early organisms by UV irradiation led to the evolution of a variety of enzyme-mediated repair systems. These include photoreactivation of dimers, dark repair of dimers, excision repair, recombinational repair, and SOS or error-prone repair. It is these repair enzymes, particularly those associated with the last three forms of repair, that pre-adapted these early prokaryotes to bacterial sex (Margulis and Sagan, 1985). Some of these repair enzymes excise thymine dimers, leaving a single stranded gap. The complementary sister strand could then be used as a template for synthesis to fill the gap. If there is no complementary strand, then the cell must either find another source for a template or die. Sometimes bacteria excrete DNA into their surroundings, often in response to UV irradiation. Often such excreted DNA is wrapped in protein coats, which make the excreted DNA more dense than the surrounding water. This is a distinct advantage if it allows these molecules to sink below the surface, and away from the high flux of UV. The other source of DNA could be plasmid or phage DNA, or homologous DNA found in solution (such DNA would be capable of transfomation- see the first section of this review on how this process may have evolved). If the source of the template is exogenous DNA, then the repair process is, by definition, a type of sexuality, because it involves DNA from more than one parent.

There are a number of lines of evidence that support this connection between the evolution of DNA repair mechanisms and the evolution of bacterial sexuality. Mutations that cause a loss of sexual competence in *E. coli* also confer extreme sensitivity to UV irradiation. Similarly, mutations which increase UV sensitivity, often by up to a hundredfold, simultaneously render *E. coli* incapable of genetic recombination. It has been shown that UV irradiation actually stimulates transformation of exogenous DNA in both *B. subtilis*, and *S. cerevisiae* (Hoelzer and Michod, 1991; Michod et al., 1988; Wojciechowski et al., 1989; Durand et al., 1993).

Even today, artifacts of this ancient battle between living organisms and UV radiation still exist. For example, oxygen is toxic to obligate anaerobes such as *Clostridium*. The ancestors of such anaerobes must have evolved prior to the appearance of the protective ozone layer approximately 1 billion years ago. If such was the case, these cells would be expected to have evolved well-developed means of defending themselves from the lethal UV flux present at that time. Interestingly, even present-day anaerobes show far more resistance to UV radiation than do aerobes.

The reasons for the maintenance of such extreme radiation resistance are unclear. Once the protective ozone layer had developed, UV irradiation posed a less serious threat, and UV repair systems may have been retained in one form or another by many organisms simply because they had come to serve more than just a repair function.

Evolution of Eukaryotes

Margulis and Sagan base much of their theory of the evolution of meiotic sex on the earlier theory of Margulis regarding the evolution of eukaryotic cells (Margulis, 1981). This theory, also known as the "Serial Endosymbiotic Theory" or S.E.T., is now generally regarded as correct, although some of its more speculative components continue to be controversial. While such an idea was not completely new, (as early as 1905, Mereschkovsky suggested that mitochondria may have at one time been free living organisms) Margulis took it a great deal farther than it had ever been taken before, developing a highly detailed theory.

Simply stated, eukaryotic cells first arose from two or more types of prokaryotic cells, which eventually came together into a composite proto-eukaryotic cell. There are two models for how such a symbiosis could begin: 1) the predator/prey model, and 2) the parasite/host model.

In the predator-prey model, a large, predatory, anaerobic prokaryote, perhaps similar to the present-day *Thermoplasma acidophila*, engulfed, but failed to digest, a smaller aerobic prey species. Although *Thermoplasma* is an archaebacterium, it nevertheless displays important eukaryotic characteristics such as a histone-like protein that surrounds and protects its DNA from degradation. The smaller aerobic prokaryotes which the *Thermoplasma*-like predator ingested became protomitochondria, and eventually evolved into mitochondria. By entering into this relationship, these two prokaryotes were able to combine the predator's resistance to thermal and pH stress with the protomitochondrion's ability to carry out oxidative phosphorylation. The combined organisms opening up for themselves a previously unavailable niche, and their local cooperation led to more effective competition in their larger niche. As photosynthetic organisms proliferated, and the concentration of atmospheric O_2 increased, this and similar mutualisms would be increasingly favored.

A similar series of events is thought to be responsible for the origin of plastids (the photosynthetic organelles found in all plants and green algae). According to this view, protists, already containing mitochondria, engulfed green *Prochloron*-like bacteria and cyanobacteria, but failed to digest all of them. The surviving *Prochloron*-like bacteria eventually evolved into the chloroplasts of the plants and green algae, while the endosymbiotic cyanobacteria evolved into the rhodoplasts of the red seaweeds.

The second model, the parasite/host hypothesis, suggests that the protomitochondria, rather than being the hapless victims of a rapacious (though not entirely successful) predator, were instead parasites of an unwilling host. In this scenario, organisms such as the aerobic *Bdellovibrio* invaded and parasitized their host. Those hosts best able to survive such an invasion passed on more of their genes to subsequent generations. Selection would eventually lead both to a less virulent parasite and a more resistant host. A symbiosis could eventually evolve favoring those protomitochondria that could both use oxygen efficiently and at the same time sequester the potentially damaging oxidative phosphorylation process from the host's DNA. Such strains would be more likely to survive than other less mutualistic members of their species.

While all mitotic organisms have nuclear membranes, not all organisms with nuclear membranes are mitotic. Margulis and Sagan suggest that this is because the nuclear membrane originated prior to the evolution of mitosis.

They suggest two additional hypotheses to account for the origin of the eukaryotic nuclear membrane. The first assumes that as genomes became larger and more complicated, selection favored an elaboration of the bacterial mesosomal membranes. Larger membranes, which were more efficient at evenly distributing the replicated DNA to progeny cells, eventually led to the evolution of what we now know as a nuclear membrane.

The second hypothesis supposes that the nuclear membrane was a response to the destructive ravages of molecular oxygen on early protist DNA. Molecular oxygen is highly reactive, and can severely damage unprotected DNA. It was toxic to early protists, and remains so today to some members of the Kingdom Protista. Because free oxygen is used in the manufacture of steroids, it has been suggested that steroids were used in proto-eukaryotes as a means of minimizing the damages to the genetic material caused by free oxygen. Later, such steroids could be utilized in the construction of membranes which were far more diverse in their properties than had been previously possible. One of the first uses for this new class of lipids was the construction of membranes enclosing what had now become the nuclear DNA.

The Evolution of the Chromosomes

Chromatin is composed of approximately 60% protein and 40% DNA. The basic organizational unit of the chromosome is a 100-angstrom diameter chromatin fibril consisting of DNA wrapped around histone and non-histone proteins. These fibrils are long thin threads which are in turn wrapped around groups of eight histones to form nucleosomes. The nucleosomes have the appearance of beads on a necklace when viewed by transmission electron microscopy.

While there is virtually no variation in the nuclear organization of members of the animal and plant kingdoms, there is a great deal of variability in the kingdom Protista. Members of this kingdom include a number of organisms which appear to be evolutionary intermediates between prokaryotes and eukaryotes. *Pelomyxa palustris* is a living relic from the

Proterozoic. This amoeba possesses a nuclear membrane, but the membrane apparently evolved prior to the appearance of either mitochondria or mitosis. It is amitotic, lacking both conventional mitotic spindles and nucleosomes. *Entamoeba* is another prototist lacking both mitotic apparatus and mitochondria. *Nanochlorum*, on the other hand, lacks mitotic apparatus but does have mitochondria.

Using the natural histories of these as well as other eukaryotic lineages as a source of inference, Margulis and Sagan conclude that in the eukaryotic line, the nuclear membrane evolved some time prior to the evolution of chromatin and chromosome-like structures. They also feel certain that the chromosomal organization of DNA evolved independently of chromosomal movement, probably some time before the divergence of the animal and plant lineages.

Microtubule Organizing Centers and the Evolution of Mitosis

Jeremy Pickett-Heaps (1971) introduced the concept of the microtubule organizing center (MTOC). Many years of microscopic observation had shown that in eukaryotic cells, a number of cell structures originated from particular locations at various times during the cell cycle. These structures included cilia, sperm tails, asters, axopods, mitotic spindles, kinetosomes, centrioles, and undulipodia. With the advent of more advanced microscopic techniques in the 1960s, it had become apparent that these locations of origin were actually centers from which the long, thin fibrous structures called microtubules were deployed (Margulis and Sagan, 1985). These microtubules, regardless of the use to which they are put, always have a constant 240 angstrom diameter. Different cellular structures have different arrangements of microtubules. For instance, cilia and flagella have a 9 + 2 arrangement of microtubules—nine pairs arranged in a circle around a central pair of microtubules. Both cilia and flagella arise from an object known as a basal body. This basal body has an arrangement of 9 triplets of microtubules surrounding a central region lacking any microtubules Interestingly, centrioles, also have a characteristic arrangement of 9 triplets of microtubules surrounding a central region lacking any microtubules.

It is clear that all of the microtubule-based structures share a common ancestry. However, there is no consensus about the nature of this ancestry. Margulis and Sagan have proposed the somewhat radical hypothesis that, like mitochondria and plastids, microtubule-based organelles are the remnants of an ancient endosymbiotic relationship. They contend that microtubule organizing centers are the stripped-down remnants of an ancient, motile spirochete which had invaded, and had later became endosymbiotic within, an ancestral prokaryote similar to *Thermoplasma*. Subsequent co-evolution stripped the endosymbiotic spirochetes (by now MTOCs) of all genetic information separate from that of their host. Genes transfered to the host genome provided them only with their ability to replicate and their penchant for contractile motion.

The motility of these MTOCs opened a whole new world to their previously torpid hosts. Not only was the MTOC's ability to contract used to propel the host through its environment by means of undulipodia, but additional copies of MTOCs were eventually co-opted for a series of complex intracellular movements, including mitosis and, eventually, meiosis. During mitosis, it is the MTOCs that are responsible for the formation of the spindle fibers, the kinetochores (also known as spindle fiber attachment sites), and the centrioles. There was strong selection for synchrony between the replication of the host's chromosomes and the replication of the remnant spirochete genomes (kinetochores). If kinetochores replicated faster than the host's chromosomes, then the chromosomes would be split apart when the kinetochore/chromosome complexes migrated towards the poles of the cell. If, on the other hand, kinetochores replicated too slowly, then one or perhaps both of the two newly formed cells would lack one or more chromosomes as well as their attendant kinetochores.

Although the model as published does not deal with this point in detail, the evolution of kinetochores would have allowed larger and more extensive genomes than were possible in bacteria to be segregated regularly. Unfortunately for the details of this story, spirochetes have no sign of α-tubulin or β-tubulin- the proteins of which all microtubules are constructed. Nonetheless, the evolution of a symbiotic relationship with a currently unknown organism having undulpodia like properties remains a possibility.

Evolution of Meiotic Sex

Meiotic sex is thought to have evolved independently a number of times in the ancestors of modern protists, perhaps through a scenario similar to the predator/prey model for the evolution of serial endosymbiosis. The primary difference between the predator/prey model for the evolution of eukaryotes and that for the evolution of meiosis is that the latter relies upon the cannibalism of conspecifics. Stressful situations such as starvation or desiccation could lead to such cannibalism. *Acanthamoebae*, as well as ciliates such as *Stentor*, have been known to cannibalize conspecifics as a response to starvation. Cleveland (1947) reported observing a case involving *Stentor*, in which attempted cannibalism resulted in the fusion of the two cells. If the engulfing organism fails to digest its prey, a doubled organism will result, containing two sets of chromosomes, two sets of MTOCs, and two nuclei, of which the last may or may not fuse.

Another means by which doubling of cellular structures could have occurred would be if cytokinesis (cytoplasmic cell division) failed following chromosomal doubling. If conditions happened to favor the doubled organism, it would outcompete its haploid equivalents regardless of the process that led to the initial duplication. Eventually, however, conditions would again favor the haploids, and there would be strong selection to return to haploidy, or perish.

Cleveland (1947) was the first to realize that kinetochore "tardyness" could relieve doubleness. That is, diploidy could be "relieved" if the kinetochores (MTOCs attached to the chromosomes) delayed replication relative to the rest of the MTOCs and DNA. If this were to occur, the net result would be that entire doubled chromosomes—rather that chromatids—would move to the poles of the dividing cell. After the kinetochores finally replicated, a second, subsequent round of cell division would split these doubled chromosomes, taking the "cannibalism and relief of diploidy cycle" full circle.

Not only was a delay in kinetochore replication a prerequisite for reduction division, but so too was the equal segregation of chromosomes into daughter cells. For this reason, it seems likely that the first, incipient meiotic cycles must have occurred in protists having low chromosome numbers. Table 2 (modified from Margulis and Sagan (1985)) shows the probabilities of survival of the progeny from cells having different numbers of chromosomes.

As can be seen, the smaller the number of chromosomes, the greater the probability of producing viable progeny.

These cycles of doubling through cannibalism, followed by relief of doubleness through tardy kinetochore replication, could, they suggest, have been driven by any of a number of naturally occurring environmental cycles. Eventually, selection could hone these processes, and add embellishments such as crossing-over and the synaptonemal complex. While many

TABLE 2. (Modified from Margulis and Sagan (1985)) Shows the Probabilities of Survival of the Progeny from Cells Having Different Numbers of Chromosomes

Number of Chromosomes	Total Number of Different Possible Progeny	Percent of Euploid Progeny
2	3	33%
4	5	20%
8	9	11%

consider crossing-over to be the primary function of meiosis, there are a number of protists that undergo meiosis, but show no evidence of crossing-over. These singular organisms undergo a one-step type of meiosis in which homologous chromosomes pair and segregate without any preliminary duplication. It remains to be determined whether they once had the ability to undergo crossing-over and lost it, or never had the capability in the first place.

The synaptonemal complex is an RNA-protein complex which holds homologous chromosomes together during prophase I of meiosis. Margulis and Sagan view this complex as a structure that originally evolved as part of a repair mechanism in diploids, allowing the repair of a damaged chromosome by the use of the undamaged homolog as a template. This repair tool was later co-opted for use during meiosis to help ensure equal segregation of homologs to daughter cells, and hence euploidy. Once the synaptonemal complex had become a common feature of meiosis in some lineages, it was a short step to use pre-existing repair enzymes to recombine the sequences of the two paired homologs during meiosis.

In stark contrast to mainstream thought about the evolution of sex, Margulis and Sagan contest the notion that sex in higher plants and animal is an adaptation at all. They proffer instead, that sex is "an imperative relic, retained because of its intimate relationship with cell function, embryological development and differentiation."

They suggest further that the protist ancestors of animals and plants were faced with a dilemma. They could either commit their MTOCs to mitotic division, or they could differentiate them into undulipodia, cilia, or other structures associated with motility, and permanently lose the ability to undergo mitosis. Their presumption that the cells must have been faced with such a choice is based on the observation that there is no known example of a motile, undulipodiated animal or plant cell that is capable of undergoing mitosis. Of course, this argument needs only one counter-example to fail.

The solution to this dilemma (if it does indeed exist) was for the differentiated, motile cell to stay in contact with the undifferentiated cell from which it originated. This, they suggest, was the beginning of animal and plant multicellularity. These early ancestors of the animals and plants had to keep at least one cell totipotent and therefore capable of mitosis and giving rise to the next generation. The cells retaining this ability were also germline cells. Thus, from the very beginning, multicellularity was inextricably linked to meiotic sex. The question of whether or not to have sex never arose. It was essentially obligatory for the ancestors of highly differentiated organisms, and for the most part remains so today. When viewed in this way, the "paradox" of sexuality evaporates. As Margulis and Sagan put it:

> It follows from our argument that the widely touted paradox, "If asexual organisms can have, on the average twice as many offspring as sexual ones, why are there not far more asexual organisms?" is not a paradox at all when sexuality is considered in its broad biological context instead of a narrowly animal one. Asexuality is rampant and sex is not necessary for the reproduction of any microbial species. . . . In all organisms other than animals and plants that do not, by definition, form embryos, asexuality is overwhelmingly prevalent. (Margulis and Sagan, 1984) pp. 19–20.

While many aspects of the Margulis and Sagan hypothesis may seem radical, even outlandish, it is worth remembering that not too long ago, the notion that organelles may once have been free living organisms, also seemed ridiculous to many.

Summary

There are some important similarities and differences between the DNA repair hypothesis and the Margulis and Sagan hypothesis for the origin of sex. One similarity is that both view recombination as having evolved specifically to repair genetic damage. Bernstein and co-workers view recombination mechanisms as having evolved specifically to repair double-stranded breaks in DNA, while Margulis and Sagan view recombination as an evolutionary response to UV-induced damage in bacteria (including double-stranded breaks).

Where the two hypotheses differ markedly is in their view on the origin of diploidy and outcrossing. According to Bernstein and co-workers, diploidy and outcrossing evolved, as the size of the genome expanded, in order to mask the increasing number of deleterious recessive mutations per genome, through genetic complementation. Margulis and Sagan on the other hand, view the question from a much more mechanical standpoint. In their opinion, diploidy evolved either as a result of conspecific cannibalism induced by starvation, or due to the failure of cytokinesis following chromosomal doubling. Margulis and Sagan never specifically address the question of outcrossing. Instead, they view sex as an inevitable response to a cyclical requirement for haploid or diploid genomes in a cyclically fluctuating environment. Once established, sex became an essential part of differentiation into multicellular organisms.

This is the most easily criticized part of the Margulis and Sagan hypothesis, because it relies on an *ad hoc* set of cyclic environmental fluctuations. The DNA repair hypothesis, with its careful consideration of genetic consequences, seems to us to provide more likely mechanisms for the evolution of diploidy and outcrossing. Nonetheless, the Margulis and Sagan view of the world remains a most valuable one. They have called our attention to the fact that there are organisms that have meiosis without crossing-over, and until other explanations for such unusual life cycles are forthcoming there remains the possibility that at least some of the reasons for the evolution of meiosis could simply be mechanical rather than genetic. The current era of genomics and genome sequencing may at last allow for a more detailed examination of the evolution of various aspects of the meiotic sexual cycle through the analysis of increasingly more accurate phylogenies.

The Contagion Hypothesis for the Origin of Sex

The contagion mechanism (Hickey, 1982; Hickey and Rose, 1988) is an interesting and persuasive hypothesis which proposes an initial short-term advantage for sex, not in terms of increased fitness of the individual undergoing sex, but rather in terms of increased fitness of "selfish" transposable elements within the individual. Hickey considers the case of a transposable element which replicates during the process of transposition. He shows that while such an element cannot spread in an asexual population except within an individual organism and its descendants, the same element will be able to spread rapidly in a diploid, outcrossing sexual population as long as it does not cause a 50% or greater decrease in its host's fitness. This is because syngamy allows an infected genome to pass the element to a previously uninfected genome and thus spread the element horizontally.

As the transposable element spreads to fixation, however, its relative fitness will decrease. At the point of fixation its fitness will be exactly equal to that of its host. Thus, as the element spreads in the population, there will be increasing selective pressure on both the host genome and the transposable element to reduce any deleterious effects of the element upon the host. As Hickey points out, the fact that a transposable element can spread to fixation in a sexual, outbreeding population, even if it is strongly selected against, makes it clear that any element in the distant past which was able to code for genes permitting meiosis and outbreeding in its host would be very likely to increase to fixation in the host population.

Bell (1993) has recently cited a great deal of evidence in support of the contagion hypothesis. He points out that in prokaryotes, the sex-like transfer of DNA between cells is mediated by parasitic genetic elements known as conjugative plasmids. These plasmids are able to spread to uninfected recipient cells by coding for the partial fusion of the infected and uninfected cells. Selection for efficient transfer has resulted in genes encoded by the plasmid which prevent fusion with other cells that carry another copy of the same plasmid. This sort of "sexual specificity" ensures the plasmid's spread to previously uninfected hosts.

Once a conjugative plasmid has infected a cell, it may either replicate autonomously or integrate into the host's genome. If it does the latter, it will be replicated along with the host's genome when the cell divides. Cells which have conjugative plasmids integrated into their genomes remain infectious. Often, when infecting a new cell, part of the host's genome will be transferred along with the parasite's DNA into the previously uninfected cell. While this sexual transfer of genomic DNA from cell to cell may play an important role in the population genetics of bacterial populations, Bell (1993) is quick to point out that "... the evolution of conjugative plasmids can be understood very straightforwardly in terms of selection for elements that are able to spread to a range of lineages, rather than being bound to a single lineage which will sooner or later become extinct."

Prokaryotes are not the only group to show evidence for a parasitic origin of sex. The vegetative segment of the life cycle of the plasmodial slime mold *Physarum polycephalum* consists of two distinct phases: a uninucleate myxamoeba, and a multinucleate syncytial plasmodium (Dee, 1982). The myxamoebae are released as spores and act as isogametes, fusing with myxamoebae of opposite mating types to form diploid zygotes which then develop into the plasmodia. In one strain examined in detail, Kawano et al. (1991) have discovered that the mitochondria fuse with high frequency during two stages of the life cycle: the diploid zygote stage, and the sporulation stage. Kawano et al. found that these mitochondrial fusing strains (mif^{+}) contain a 16 k.b. plasmid, which encodes for the ability of mitochondria to fuse, and also for fusion between mitochondria and nuclei. This mt-nuclear fusion is followed by recombination between homologous mtDNAs, or between mitochondrial and plasmid DNA, which is in turn followed by several mitochondrial divisions (accompanied by mt-nuclear divisions). The end result is recombinant mtDNA, while the plasmids are then passed on intact to all mitochondria. As these authors put it: "The processes of mt-nuclear fusion, recombination and segregation may be regarded as providing a kind of mitochondrial 'meiotic' cycle."

Bell (1993) points out that the mating type "alleles" of all of the bipolar fungi studied so far are not alleles at all, but rather 'idiomorphs'. The term "idiomorph" has been coined for genes occupying homologous positions on the chromosome but having such dissimilar sequences that they cannot be of homologous origin (Metzenberg and Glass, 1990). Bell refers to these idiomorphic mating type alleles as "ancient infections." These parasitic sequences or "infections" have since been co-opted by the cell, and are now thoroughly integrated into the cell's life cycle.

Bell proposes one possible scenario to explain these observations. Suppose a parasitic genetic element similar to the bacterial conjugation plasmids (perhaps a transposon) were to infect a eukaryotic lineage. By coding for the fusion of infected cells with uninfected cells, such an element would quickly spread throughout the asexual members of the lineage, converting them to infectious, sexual cells. This element would spread even if it lowered the fitnesses of its hosts considerably. Such elements may become integrated into their hosts' genomes. As mentioned earlier, selection would favor sexual specificity so that infected cells are less likely to fuse with previously infected cells.

Once such an element had spread throughout the entire population, it would, in essence, turn the population back into an asexual lineage (there would no longer be any uninfected cells to fuse with). Once this had occurred, a second, different genetic parasite might infect this lineage, and spread through it. It is important to note that in such a situation, the two genetic elements would behave antagonistically toward each other with respect to cell fusion. The element which had spread to fixation would act to prevent further fusion because of sexual specificity, while the new genetic element would code for fusion with all cells lacking copies of itself. Bell points out that such a conflict could be reduced if the second invading element were to come to occupy a position on the chromosome that is homologous to the first element.

If the sexual parasites were transposable elements, it is quite possible that they may occasionally have occupied the same locus on homologous chromosomes. The yeast *Saccharomyces cerevisiae* has idiomorphic mating type alleles *MATa* and *MAT*α which behave in a manner remarkably similar to transposable elements. Haploid cells of this yeast can sometimes switch mating types when there are no cells of the opposite mating type with which to mate. This mating type switching occurs through a transposition event, in which *MAT* information is transposed from one of two silent storage loci to the expressed *MAT* locus. Additional evidence to support Bell's contention lies in the fact that in the very distantly related yeast *Schizosaccharomyces pombe*, the *MAT* alleles can loop out to form a plasmid which can subsequently reintegrate into the chromosome at the appropriate position (Beach and Klar, 1984).

Bell (1993) also points out that in fungi in which the genes controlling sexual and asexual phases are separate, the sexual genes are idiomorphic, while the genes controlling the asexual phase are allelic. In those fungi in which both sexual and asexual phases are controlled by the mating type locus, the presence of different idiomorphs is required for sexual fusion, while only cells carrying the same idiomorph can fuse somatically. Bell suggests that this is to be expected if idiomorphs are in fact ancient parasites.

He goes on to explain that this is due to the fact that such parasites should "attempt to infect lineages from which they were previously absent but to pool cytoplasmic resources with individuals who are already infected." Such resource pooling could be considered a form of kin-selection.

Michod and Long (1995) have pointed out some intimate relationships between the DNA repair hypothesis and the contagion hypothesis, in which each postulated mechanism can benefit from the existence of the other. They model a situation in which damaged sexual cells are allowed to mate with asexual cells, most of which are undamaged, and find that this gives a great advantage to the sexual cells even if they are rare. They point out further that the most realistic situation under which such a seemingly unexpected sort of mating might occur is through contagion. A contagious element that permits fusion and recombination, such as an F factor in bacteria, will produce exactly the sort of situation that they model.

Recently, an experimental test of one aspect of the contagion hypothesis was carried out by Zeyl et al. (1996). They followed the spread of a retrotransposon, Ty3, in diploid, and thus effectively asexual, populations of *S. cerevisiae*, and in sexual populations that they formed by mixing together equal numbers of *MATa* and $MA^{T}\alpha$ cells. Their reasoning was that, even though the element tends to be deleterious for a variety of reasons, it might spread more readily in a sexual population than an asexual one through horizontal transfer.

This prediction was borne out. Although the element did sometimes spread in an asexual population, apparently because of unrelated selective events, it consistently reached higher levels in the sexual populations.

Summary

In summary, the contagion hypothesis suggests that ancient genetic parasites, similar to present-day conjugative plasmids or transposable elements, first infected and then spread horizontally through a number of eukaryotic lineages. As long as these parasitic sexual elements could be transferred horizontally, they could exact a heavy toll on their hosts and still manage to spread. Once they had spread to fixation, however, selection would act swiftly to reduce their virulence. Those elements which did not lose their virulence would drive their hosts along with themselves to extinction. Subsequent invasions by different parasitic sexual elements would be followed by coevolutionary interactions that would eventually lead to the numerous types of mating systems seen today, particularly if the new elements integrated into the genome as idiomorphic alleles of the earlier elements. One very great advantage of this hypothesis is that it explains the very different idiomorphs that have been found to control mating in different fungal species. The connection to other sexual systems is less obvious but can still be made with some effort.

EXPERIMENTAL EVIDENCE FROM ISOGENIC SEXUAL AND ASEXUAL STRAINS OF YEAST

We have examined the advantages of the ability to carry out sexual recombination, using the yeast *Saccharomyces cerevisiae* (Birdsell and Wills, 1996). *Saccharomyces* has the advantage that its ability to undergo recombination is controlled by two idiomorphic alleles, *MATa* and *MAT*α, that can be switched by a transposition mechanism. *Mat a/*α heterozygotes are sexual and are capable of undergoing meiotic recombination while *Mat* homozygotes are not. In addition to the ability for meiotic recombination, *Mat* heterozygotes have greatly elevated mitotic recombination rates, DNA repair capabilities, and transformation abilities.

We established pairs of otherwise genetically identical diploid lines that differed only in being homozygous or heterozygous at *MAT*, and determined which of these would win in serial transfer competition experiments. Because no marker genes were employed to distinguish between the pairs of strains (they were distinguished by a phenotypic assay on plates), we could be confident that any differences seen between them could be traced to the effects of the *MAT* locus.

It was found that the *MATa/*α strains would always win, but that they would win significantly more quickly if the genetic background was heterozygous and the sexual strain was allowed to undergo a round of meiotic recombination prior to the competition. Presumably, meiotic recombination shuffled the genetic variation present producing progeny with a much higher variance in fitness than the parental generation. It appears that at least some of these progeny had a much higher fitness than the mean fitness before recombination, and it was these progeny that were able to more rapidly outcompete the non-recombining asexual strain. When the genetic background of the competing isogenic sexual and asexual strains was homozygous, a round of recombination actually initially reduced the sexual strain's relative competitive ability. The disadvantage of recombination in a homozygous background may be due to the fact that in the absence of genetic variation, there is no benefit to recombination, while there is a cost- the cost being that recombination itself, with its extensive breakage and repair of DNA during crossingover, is mutagenic. Recombination has always been assumed to be an error free pathway, however, in the early 1960's it was found that meiosis actually induced mutation in yeast (Magni and von Borstel, 1962; Mangi, 1963). This rather amazing finding seems to have been ignored by evolutionary modelers, which, to our knowledge have never incorporated recombinational mutagenesis into models of the evolution of recombination. These early findings have since received extensive support from a series of elegant experiments conducted by Strathern et al. (1995), Holbeck and Strathern (1997), Holbeck and Strathern (1999) and McGill et al. (1998), which demonstrated that recombination increased the mutation rate at a nearby reporter gene by over 100 fold.

In our competition experiments, the largest effect on competitive ability was conferred by heterozygosity at *MAT per se*, whether or not the sexual member of the competing pair had been allowed to undergo a round of recombination. This may be attributable in part to the many differences in DNA repair between *MAT* heterozygous and homozygous strains, as well as to the increased rates of mitotic recombination in the *Mat* heterozygotes.

A more subtle effect could be seen by following the *MAT* homozygous and heterozygous strains during the course of the competitions. When the genetic background was homozygous, the *MAT*-heterozygous strains actually increased in their relative fitness during the competitions. We have shown more recently (Selk and Wills, 1997) that this is primarily due to the accumulation of harmful mutations of small effect in *MAT*-homozygous lines, mutations which are probably taking place in the *MAT*-heterozygous lines as well but are being repaired at a higher rate.

These results suggest that there are multiple sex related advantages to heterozygosity for these idiomorphic alleles. These advantages encompasses several aspects of DNA repair, lending support to the idea that such repair played an important role in the early evolution of sex. The advantages include an increased ability to transform exogenous DNA, which may tie in with either the nutritional or DNA repair hypotheses for the evolution of transformation. If any of these factors were playing a role as the original vectors carrying the mating idiomorphs spread through the population of yeast-ancestors, then it can be easily seen why such vectors could have a substantial advantage.

THE MAINTENANCE OF SEXUAL RECOMBINATION

An important point which has not always been clearly stated in discussions of the evolution of sex is the distinction between the *origin* of sex and its *maintenance* once it has arisen. The processes which brought about the origin of sex are not necessarily the same processes responsible for its maintenance. In particular, once heterogamy and outbreeding have evolved, then sex in such organisms involves a twofold cost (Maynard Smith, 1978). This cost is due to the fact that in any outbreeding population of a fixed size, any parthenogenetic female that appears will be able to produce twice the number of females that a sexual female can. Parthenogenesis can spread very rapidly through such a population. This means that whatever forces are maintaining sex in such a population, they must be able to do so in spite of its twofold cost. However, for most of the organisms that we have considered in the first section of this review, the twofold cost is not a difficulty. It seems likely to have presented a problem only fairly recently in the history of life.

Once sex had evolved, other processes could have taken over to maintain it. These processes need not be, and in all likelihood are not, a completely different subset from those already considered. For example, both parasite-host coevolution, DNA repair and the concealment of the effects of deleterious mutations can provide strong selective pressure for the maintenance of sex. In the second half of the review, we consider other possible mechanisms which have in general been investigated theoretically to much greater depth than the mechanisms involved in the origin of sex that were dealt with in the first half.

I. Maintenance of Sex Models which Involve the Removal of Linkage Disequilibrium

A.1. "Vicar of Bray" or "Fisher-Muller" Type Models which Invoke Mutation as the Driving Force behind the Evolution of Sex

Both Fisher (1930) and Muller (1932) are usually credited with the idea that amphimixis (outcrossing) would be selected for because it is the quickest way in which two independently arising *advantageous* mutations, occurring in a finite population, could come to be found in the same individual (Felsenstein, 1988). Simply stated, this model views sexual recombination as allowing evolution to proceed more rapidly. It has been pointed out, however (Bell, 1982; Kondrashov, 1993), that while Fisher and Muller are traditionally given credit for this hypothesis, Morgan (1913), Guenther (1906), and even Weismann (1889) all proposed essentially the same idea years earlier.

Bell (1988a) provides a simple scenario (as have others) with which to illustrate the point. Suppose that there are two haploid populations that have been living in a stable environment for a long time. With the exception of an occasional rare mutation, all loci are fixed for the optimal allele for this environment. One population is sexual, the other asexual. Now suppose that the environment suddenly changes such that the original genotype is no longer optimal. Suppose further that two rather rare mutations can arise that are strongly favored in this new environment. One of these mutations can occur at locus A and the other can occur at an unlinked locus B. In the asexual population, it will usually be a very long time before both of these mutations are found together in the same individual. The most likely situation is that one of the mutations will be driven to a sufficiently high frequency that there is a reasonable chance for the second mutation to occur in an individual that already carries the first. Without sex and recombination, these two mutations are much more likely to be found in separate cells, and would therefore be in competition with each other. In the sexual population, on the other hand, the two favorable mutations can be brought together in one individual through sexual recombination. Bell (1982) has dubbed this theory the "Vicar of Bray" model after an English cleric noted for his mercurial habit of changing his religion to that of the newly crowned monarch whenever the throne changed hands. Others, including Felsenstein (1988) refer to it as the "'Fisher-Muller" model. We will use the former term as it eliminates the possibility of confusing this model with the somewhat similar-sounding "Muller's ratchet" (see below).

An important point regarding the Vicar of Bray hypothesis, which should be made clear from the start, is that it is a group selection theory, "... the advantage invoked is an advantage that accrues to populations or species, rather than to individuals ..." (Bell, 1982).

Felsenstein (1988) points out that the Vicar of Bray hypothesis relies on linkage disequilibrium which arises through random genetic drift in a finite population. He goes on to give the following example:

"... if two favorable mutants happen to occur in the same generation in a *diploid* population of size N, one at the A locus and one at the B locus, then the chance that these will occur in different gametes is $[1 - 1/(2N)]$. They will occur in the same gamete $1/(2N)$ of the time. Either way, the population will be in linkage disequilibrium. The disequilibrium is random and a result of the finiteness of the population size. If the population were of infinite size, each mutant would recur many times and would arise at a frequency μ. A fraction μ^2 of the gametes would be double mutants, so that the population would be at linkage equilibrium."

The first attempts to quantify this model were made by Muller (1958; 1964) and by Crow and Kimura (1965). The latter authors attempted to compare the rate at which favorable gene combinations could be incorporated into a population in sexual and asexual lineages. They began with the observation that while sexual recombination may be advantageous because it can produce many different, new genotypes from a few genes, it also breaks up these new combinations just as readily as it creates them (the so-called recombinational and segregational loads).

They go on to point out that for a given amount of genetic variation, selection is more efficacious in an asexual population than in a sexual one. This is because in the former it can act on the total genotypic variance rather than just on its additive component.

While selection may be more efficient in an asexual population, there is a limit to how far selection can go in such a population. This limit is set by the fittest genotype. A sexual population, on the other hand, has no such preset limitation.

Crow and Kimura's Model In Crow and Kimura's paper (1965), the variables N, U, g and s were defined as follows:

N = the population size

U = the total rate of occurrence per individual per generation of favorable mutations at all loci.

g = the average number of generations between the occurrence of a favorable mutation and the occurrence of another favorable mutation in a descendant of the first.

s = the average selective advantage of a favorable mutant.

In a population in which only beneficial mutations are considered, and in which the selection coefficient s for such mutations is small, Crow and Kimura found that "the ratio of incorporated mutations in a sexual population to that in an asexual population is NUg :1" (see Table 3). They go on to conclude that "the advantage of a reproductive system that permits free recombination is greatest for the incorporation of mutant genes with individually small effects, occurring at relatively high rates, and in a large population."

As can be seen from the table, sexual recombination can have a very great advantage indeed under certain conditions. As either N, or the ratio U/s, increase, evolution proceeds with increasing rapidity in sexual populations relative to asexual populations. As will be seen below, however, this advantage is strongly dependent on the assumptions that underlie the model.

The second situation that Crow and Kimura considered was one in which the individual mutations were by themselves deleterious, but when combined they were advantageous. Under such conditions, once the double mutant AB arose in an asexual line, it would become fixed at a rate

TABLE 3. The Ratio of the Rate of Incorporation of New Beneficial Mutations into a Sexual Population to the Rate of Incorporation into an Asexual Population (Crow and Kimura, 1965)

	N						
U/s	10^3	10^4	10^5	10^6	10^7	10^8	10^9
10^{-7}	1.0007	1.01	1.12	2.38	16.7	162	1.63×10^3
10^{-6}	1.007	1.09	2.15	14.4	139	1.4×10^3	1.4×10^4
10^{-5}	1.07	1.92	12.1	116	1.2×10^3	1.2×10^4	1.2×10^5
10^{-4}	1.69	9.75	92.6	922	9.2×10^3	9.2×10^4	9.2×10^5
10^{-3}	7.5	69.6	691	6.9×10^3	6.9×10^4	6.9×10^5	6.9×10^6
10^{-2}	46.7	462	4.6×10^3	4.6×10^4	4.6×10^5	4.6×10^6	4.6×10^7
10^{-1}	240	2.4×10^3	2.4×10^4	2.4×10^5	2.4×10^6	2.4×10^7	2.4×10^8

determined by its selective advantage over the wild type ab. They concluded however that, assuming unlinked loci, sexual recombination would be at a "distinct disadvantage" due to the fact that in order to get from the predominating ab genotype to the more fit AB genotype, a sexual population must pass through an "adaptive valley" of ill-fit Ab and aB genotypes. Such a valley is similar to those envisioned by Wright in his adaptive landscape model (Wright, 1932; Wright, 1988). One way in which a sexual population could cross such a valley would be for it to drift across the valley as the result of a population size bottleneck. Kimura (unpublished) showed, however, that even for an effective population size of only 1,000 individuals, the probability of the joint fixation of two mutations, each having a selective disadvantage of 1%, with a 5% advantage when combined, is only $2.5p_0q_0 \times 10^{-6}$, where p_0 and q_0 are the initial allele frequencies for the two mutants. Thus Crow and Kimura (1965) conclude, "if this type of gene action were the limiting factor in evolution at the time sexual reproduction first evolved, sexual recombination might never have been 'invented'."

When they re-examined the situation in which individual mutations were by themselves deleterious, but advantageous when combined, this time allowing for genetic linkage between the newly arisen mutations, they found that the double mutant AB should increase and become fixed in a sexual population provided that $s > r$ (where r is the recombination frequency between the two linked genes, and s is the selective advantage of the double mutant over the wildtype). Using Wright's analogy, "the effect of linkage is to raise the valley between the two adaptive peaks and with extremely close linkage to provide a direct bridge."

They conclude:

> ". . . the development of sexual reproduction confers no immediate advantage on the individual in which it occurs. In fact, the result is far more likely to be deleterious. The benefit is only to the descendants, perhaps quite remote descendants, and to the population as a whole. Thus, it seems likely that the selective mechanism by which recombination was established was intergroup selection."

Maynard Smith (1968) was the first to point out that there was a flaw in of Crow and Kimura's assumptions. He did this by giving a counterexample for which, according to Crow and Kimura's reasoning, one would expect sex to confer a distinct advantage, and was able to show that sexual reproduction showed no advantage at all over asexual reproduction. The counterexample consisted of a case in which a change in the environment of a haploid organism would favor evolution proceeding from ab to AB. He assumed that fitnesses were multiplicative, the population was at mutation-selection equilibrium before the change in environment (i.e.: freq.(AB)freq.(ab) = freq.(Ab)freq.(aB)), and that genetic variation was being maintained by recurrent mutation. He then showed that, following the change in environment, sexual recombination would *not* accelerate the rate of evolution, "all that sexual reproduction can do is to restore this independence relationship." The discrepancy between his conclusions and those arrived at by Kimura and Crow was based on the fact that the latter viewed mutations as being *unique* events, while in Maynard Smith's model, mutations were viewed as being *recurrent* events that have occurred sufficiently frequently that they have reached linkage equilibrium. He pointed out that the statement by Crow and Kimura that NUg beneficial mutations will be incorporated by a sexual population every g generations makes the implicit assumption that all these mutations are unique. To the extent that such is not the case, Crow and Kimura's model will overestimate the rate at which sex accelerates evolution. Their estimate is "out by a factor equal to the number of times a particular mutation recurs in g generation{s} in a population of N individuals."

Not only will the uniqueness assumption overestimate the rate at which sex accelerates evolution, but it will also underestimate the rate at which asexual clones evolve (Maynard Smith, 1968). This is due to the fact that it neglects the possibility of more than one individual, in any given generation, giving rise to a particular favorable mutation. Instead, it assumes that the first individual to have a favorable mutation in any given generation is the only individual to have any descendants many generations later. A favorable mutation rate of, say, 10^{-8} is by no means synonymous with "uniqueness", especially when one is concerned with the evolution of sex in microorganisms which regularly have population sizes in excess of 10^{9}.

If one assumes for the moment that Maynard Smith is correct in his assessment of Crow and Kimura's model, then one would expect their overestimate of the advantage of sex to be the *greatest* when the population size N is large, the mutation rate U is high, and s is small (and therefore g is large). Maynard Smith is quick to point out that such conditions are exactly those listed by Crow and Kimura (1965) as being those under which sexual reproduction is *most* advantageous.

In his conclusion Maynard Smith re-emphasizes the importance of the initial assumptions of a model in determining what conclusions will be drawn. In his counterexample, the genetic variation of the population was generated by recurrent mutation, and in this case, sexual recombination was shown to have no effect upon the rate of evolution. If, however, one were to assume that the genetic variation in the population had been generated by selection favoring different genotypes in different environments, then, sexual recombination could accelerate adaptation to a *new* environment by bringing together these co-adapted gene complexes.

> "... In other words, sexual processes are an advantage because they make it possible to bring together in one individual, not merely mutations which have occurred in different ancestors (because the same result can be achieved equally well by recurrent mutation), but different regions of DNA which have been programmed by natural selection in different ancestral populations in different environments." (Maynard Smith, 1968)

Crow and Kimura (1969) wasted no time in responding to Maynard Smith's criticism. They admitted that his counterexample did "invalidate a very minor part of Muller's original argument," but not the essential elements of it. Muller himself thought that asexual reproduction would suffer a disadvantage with respect to sexual reproduction. This is because, with clonal reproduction, in order for a favored genotype to become "fixed," it must compete against and eliminate all other advantageous genotypes (Muller, 1932). This interclonal competition would presumably slow down evolution of asexual organisms.

With sexual recombination, however, it was likely that the different advantageous mutants could "interpenetrate and cooperate" (Crow and Kimura, 1969). It was not until 1964 that Muller was convinced by an argument virtually identical to that presented by Maynard Smith (1968), that the competition between genotypes was actually the same in both sexual and asexual populations. Muller was even reportedly "relieved" that what he now called the "competition hoax" was resolved. He had hoped to write an article to this effect, but died before this could be accomplished (Crow and Kimura, 1969). For these reasons, Crow and Kimura had felt it unnecessary to mention in their original paper the point that had been brought up by Maynard Smith. They reiterated the main premise of their 1965 article, the point that Muller had made, that the importance of sexual recombination lay in the combining together, into a single individual, of rare advantageous mutations that had occurred in different individuals. They went on to point out, as had Fisher (1930), that the rate at which alleles are substituted in a sexual species is greater than the rate of substitution in an asexual species by a factor approximately equal to the number of simultaneously evolving loci (Crow and Kimura, 1969). Where they differed from Maynard Smith was in their belief that the *advantageous* mutation rate is so low that even in a very large population there would be no double mutants. They stated however that they had made no assumption about the mutations being unique, only "very rare."

At the end of their reply to Maynard Smith, Crow and Kimura conclude that the merit of their argument versus that of Maynard Smith's depends upon whether evolution proceeds primarily through the incorporation of rare, advantageous mutations or through a "shifting balance of gene combinations where individual alleles are at intermediate frequencies."

Following this rebuttal, Maynard Smith (1971) admitted that there was a "serious drawback" to his model. The drawback was that prior to the environmental change favoring the rare mutants, the frequency of the single mutants may be as low as 10^{-8}, and the frequency of the double mutant as low as 10^{-16}. It can be seen therefore, that in real populations, the double mutant would almost certainly be totally absent! Maynard Smith (1971) went on to develop another model with which to analyze the evolution of sex in terms of the Vicar of Bray theory. In this paper, Maynard Smith compared the rates of evolution between asexual and sexual haploid populations which were initially homogeneous genetically (in his previous model, the populations had been at mutation-selection balance prior to the environmental fluctuation). As before, N is population size. Favorable mutations were allowed to accumulate at any of L loci, with a mutation rate of μ per locus per generation, each favorable mutation having a selective advantage s. Fitnesses were assumed to be multiplicative. Using this new model, Maynard Smith made calculations for the following values of parameters: $\mu = 10^{-9}$, $s = 10^{-2}$, $L = 10$, and $N = 10^6, 10^7, 10^8, 10^9$, and 10^{10}. Two runs were made for each calculation in order to give an indication of the importance of stochastiscity. The results of these calculations are given in Table 4.

He found that in a population of $N = 10^6$, recombination will have no effect upon the rate of adaptive evolution. However, as the size of the population increases, so does the speed at which the sexual population can evolve in relation to the asexual. Between population sizes of about 10^8 and

TABLE 4. Relative Rates of Evolution for Sexual and Asexual Populations

	N						
U/s	10^3	10^4	10^5	10^6	10^7	10^8	10^9
10^{-7}	1.0007	1.01	1.12	2.38	16.7	162	1.63×10^3
10^{-6}	1.007	1.09	2.15	14.4	139	1.4×10^3	1.4×10^4
10^{-5}	1.07	1.92	12.1	116	1.2×10^3	1.2×10^4	1.2×10^5
10^{-4}	1.69	9.75	92.6	922	9.2×10^3	9.2×10^4	9.2×10^5
10^{-3}	7.5	69.6	691	6.9×10^3	6.9×10^4	6.9×10^5	6.9×10^6
10^{-2}	46.7	462	4.6×10^3	4.6×10^4	4.6×10^5	4.6×10^6	4.6×10^7
10^{-1}	240	2.4×10^3	2.4×10^4	2.4×10^5	2.4×10^6	2.4×10^7	2.4×10^8

Number of generations before the average number of favorable alleles per individual is either 5 or 9.9, out of a total possible of 10, for sexual and asexual populations. N = population size (Modified from Maynard Smith (1971)).

10^{10}, the sexual population can evolve at about five times the rate of the asexual (Maynard Smith, 1971).

To understand these results, it is useful to refer to Figs. 1–4, which are modified from those originally drawn by Muller (1932).

These figures illustrate the importance of population size in determining whether recombination will or will not accelerate the spread of favorable mutations. In a large population, many *different* favorable mutations may occur in different individuals, and may exist in the population at the

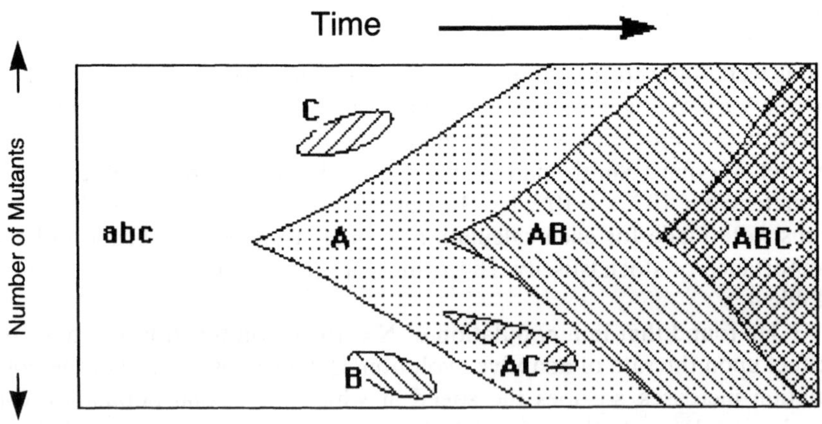

FIG. 1. Fate of Adaptive Mutations in a Large Asexual Population.

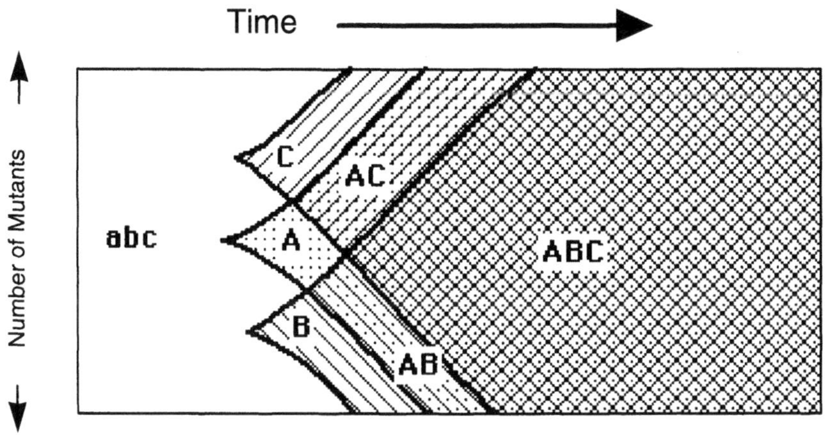

FIG. 2. Fate of Adaptive Mutations in a Large Sexual Population.

same time. Without recombination, these different favorable mutations will remain in separate individuals, and may actually compete against each other. No double mutants will occur for quite some time due to the fact that the number of individuals having one of the favorable mutations is, at least in the beginning, very small. If, on the other hand, sexual recombination occurs, then, conceivably within one generation double mutant individuals could be produced by sexual recombination between single mutant parents. Thus, in a large population, sexual recombination can greatly accelerate the spread of advantageous mutations.

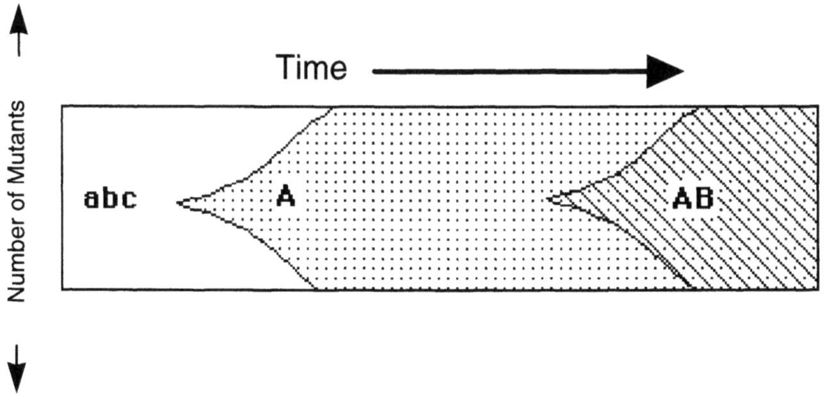

FIG. 3. Fate of Adaptive Mutations in a Small Asexual Population.

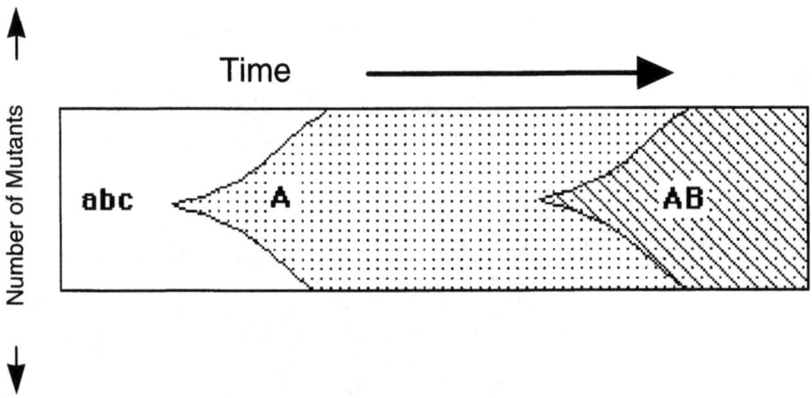

FIG. 4. Fate of Adaptive Mutations in a Small Sexual Population.

The story is quite different in a small population. In this case, the length of time between the occurrence of one favorable mutation and that of another will, on average, be very long since at any one time there will be very few individuals in which such mutations can occur. Under these circumstances, any favorable mutation that does become established in the population will spread to fixation long before another favorable mutation arises, and therefore recombination will play no role in determining the rate at which mutations become fixed in a small population.

Two other diagrams (Figs. 5 and 6), redrawn from Maynard Smith (1978) by Bell (1982), will help to illustrate this important phenomenon further. In these generalized figures, there are two critical time intervals; T_e (the length of time between the establishment of one new, favorable mutation, and the establishment of a subsequent new, favorable mutation) and T_s (the time it takes a newly established mutation to spread). During the establishment phase, the mutation is at a low frequency, and is therefore subject to predominantly stochastic effects as a result of random sampling. Once established, however, the new allele reaches a sufficiently high frequency that selection can take over, and the mutation spreads deterministically to fixation (Bell, 1982). In this model, the effects of amphimixis (outcrossing) depend critically upon the relative lengths of T_e and T_s, which in turn depend upon the magnitudes of N, μ and s. If the length of time it takes a new mutation to spread is less that the time it takes for it to become established ($T_s \ll T_e$), then amphimixis cannot accelerate evolution (Fig. 5). This is because, once a new mutation arises, it will already have become

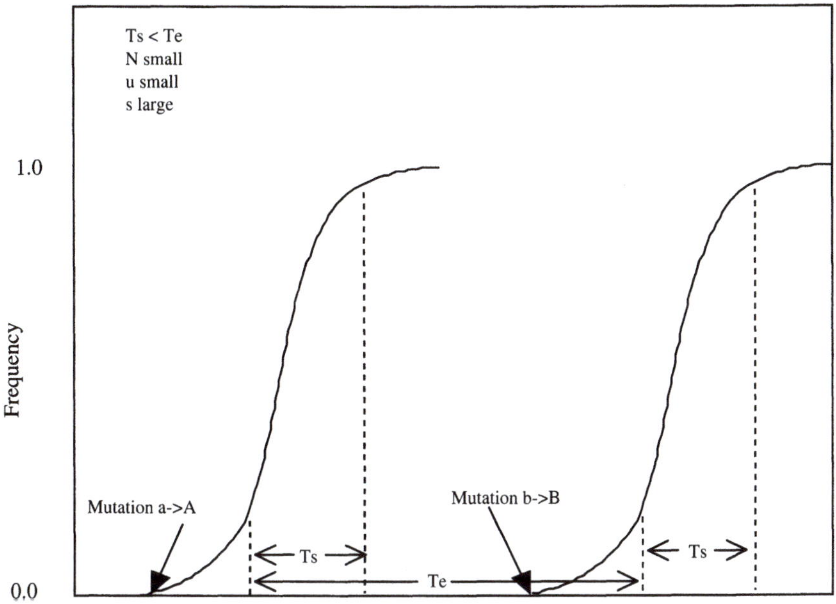

FIG. 5. Fixation of adaptive mutations under conditions in which the establishment of a new mutation takes much longer than its subsequent spread to fixation.

fixed before the next new mutation occurs, and therefore there is no advantage to recombination. Such conditions are likely to exist when N is small, μ is small, and s is large. In other words, sex would be of no advantage in small populations exposed to strong selection.

If on the other hand $T_s >> T_e$, then recombination can greatly accelerate the rate of evolution of the amphimictic population relative to the apomictic one (Fig. 6). Under such conditions, due to the relatively slow rate of spread of favorable mutations after they arise, it is very likely that the previous favorable mutation will still be at a low frequency when the next favorable mutation occurs. This makes it unlikely that the next favorable mutation will occur in an individual possessing the previous favorable mutation, and therefore amphimixis can greatly accelerate the fixation of the favorable double mutant. The conditions required for $T_s >> T_e$ exist when N is large, μ is large, and s is small. Thus amphimixis is most advantageous in large populations exposed to mild selection.

Maynard Smith concluded that Fisher (1930) was correct when he stated that "the comparative rates of progress of sexual and asexual groups . . . are therefore dependent upon the number of different loci which are

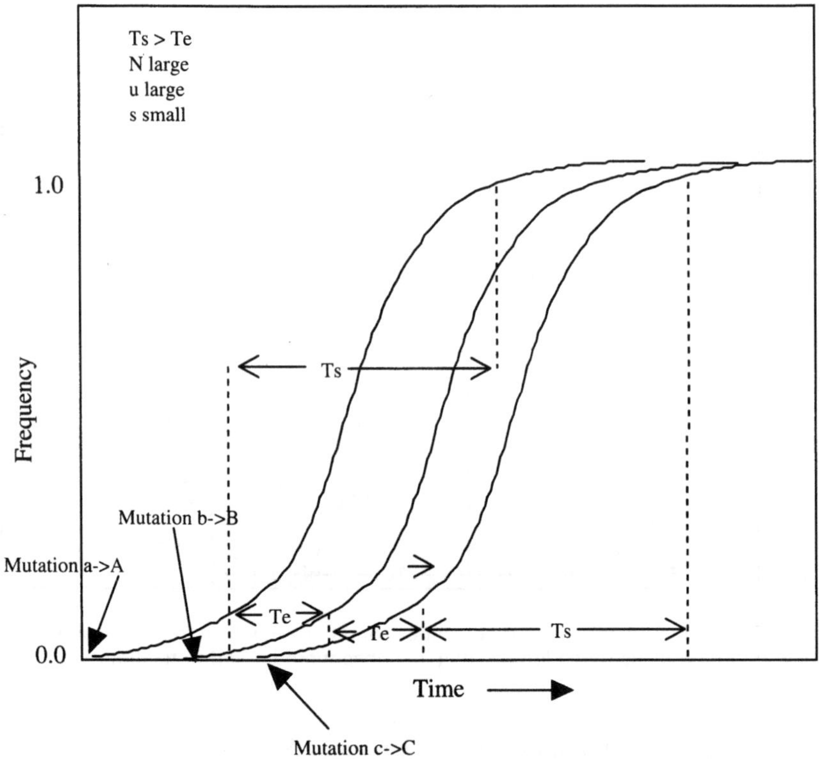

FIG. 6. Fixation of adaptive mutations under conditions in which the establishment of a new mutation takes much less time than its subsequent spread to fixation.

freely interchangeable". Under the assumptions of this model, while sex will not accelerate the spread of advantageous mutations in small populations, in large populations (in which $T_s >> T_e$), sex will accelerate evolution by a factor comparable to the number of loci (L) at which advantageous mutations can occur (Maynard Smith, 1971).

Williams (1975) is to be credited with pointing out an inherent flaw in the logic of the Vicar of Bray hypothesis. As we have seen, according to this hypothesis, the only time that sex is advantageous is when it occurs within very large populations. However, the proposed advantage (of accelerated evolution) is one of group selection. Because, if it is to be effective, group selection relies upon a population being subdivided into many semi-isolated groups, these two attributes are in direct conflict with each other.

Up until this point, most of the models discussed have assumed that fitnesses are multiplicative, that is, that the fitness of the double mutant is equal to the product of the fitnesses of the single mutants. Following the debate between Crow and Kimura on the one hand, and Maynard Smith on the other, Eshel and Feldman (1970) jumped into the fray, making an attempt to resolve the Vicar of Bray question once and for all by rigorously analyzing a situation similar to that studied by Crow and Kimura (1965; 1969), and Maynard Smith (1968). In their model, Eshel and Feldman evaluated the effects of recombination on an infinite, panmictic haploid population with constant fitnesses, consisting of individuals having two loci (a and b), in which mutation proceeded from a to A and from b to B. Unlike Maynard Smith, however, they did not consider only multiplicative fitnesses. Eshel and Feldman found that the situation was much more complicated than previously appreciated, and that the outcome depended critically upon the initial assumptions one makes regarding the fitness interactions of mutations. While the traditionally held view was that recombination enables favorable, cooperative, double mutants to appear more quickly and in higher frequency than clonal reproduction, they found that exactly the opposite was true. For four possible cases, they obtained the following results:

1. In the "supermultiplicative" case, in which the fitness of the double mutant is greater than the product of the fitnesses of the single mutants (i.e.: $w_{AB} > w_{aB}w_{Ab} > 1$), they found that after the first generation, the frequency of the favorable double mutant was always higher in the asexual than in the sexual population.
2. In the "submultiplicative" case it was found that, when the double mutant is more advantageous than either single mutant, but not as advantageous as the product of the single mutants (i.e.: $w_{Ab}w_{aB} > w_{AB} > (w_{Ab}$ *and* $w_{aB}) > 1$), after the first generation, the sexual population always had more of the double mutants than the asexual.
3. They then considered the case in which each of the single mutants is deleterious (but not too deleterious), and the double mutant is favorable (but not too favorable) ($w_{AB} > 1 > (w_{Ab}$ and $w_{aB}) > w_{Ab}w_{aB} > (1\text{-}r)w_{AB}$ where r is the recombination fraction, and where $w_{Ab} > 0.5$ and $w_{aB} > 0.5$). Assuming that mutation is "sufficiently rare", they showed that with recombination, the favorable double mutant cannot increase to fixation (actually the frequency of the double mutant cannot exceed a value which approximates the mutation rate), whereas in an asexual population, the double mutant will proceed to fixation.
4. In the special case analyzed by Maynard Smith (1968), in which fitnesses are strictly multiplicative ($w_{AB} = w_{Ab}w_{aB}$), Eshel and Feldman confirmed Maynard Smith's results, showing that they hold true

even when the fitnesses of the single mutants are not equal ($w_{Ab} \neq w_{aB}$). This is due to the fact that if one starts with an infinite population of AB individuals, with an initial linkage disequilibrium $D_{(t=0)} = 0$, the linkage disequilibrium will continue to equal zero in both the sexual and asexual populations for every subsequent generation.

Thus, if the fitness advantage of a double mutant depends at all on positive epistasis, then amphimixis will retard rather than accelerate evolution. In the discussion section of their paper, Eshel and Feldman conclude that, in spite of their results, recombination is ubiquitous in nature, and must therefore play some critical evolutionary role. They concede that their analysis applies *only* to very large, panmictic populations living in unchanging environments, and they make the interesting observation that

> "... the disadvantage of recombination in slowing the progress of the double mutant may become an advantage in the case of a changing environment. Rather than hastening the fixation [of] the double mutant, the advantage of recombination may be in prolonging the polymorphic state."

This view seems to foreshadow a similar view expressed by Hamilton et al. (1990) some twenty years later, who viewed sex as an adaptation that "stores genes that are currently bad but have promise for reuse at some time in the future."

Thus, resolution of even these relatively small aspects of the evolution of sex can be a very complicated, time-consuming process. Just as one thinks, finally, that the case has been solved, someone points out something else which seems to bring the whole question back to square one. As can be seen by the series of scientific exchanges summarized here, the outcome of these various models depends critically upon the initial assumptions imposed by the modelers.

Muller's Ratchet Muller (1964) was the first to point out that asexual populations incorporate a kind of ratchet mechanism. Once harmful mutations begin to arise in an asexual lineage, that lineage can never contain fewer harmful mutations than it contains in its least loaded line at a given point in time. The only way in which such a lineage could contain fewer mutations would be through back mutation, which can usually be ignored. Compensatory mutation may help save a population from some of the effects of the ratchet, however there is a limit as to how long compensatory mutation can forestall the ratchet. An important point regarding Muller's ratchet is that it relies upon a *finite* (though not necessarily small) population size. The reason for this is that, at equilibrium between mutation and selection, there will be classes of individuals having different numbers of mutations, 0, 1, 2, 3, 4 . . . At such an equilibrium, in a finite population, the number of individuals having zero mutations may be quite small (Maynard Smith, 1988). Eventually, the individuals in this class may not leave any offspring simply by chance. Alternatively, the individuals, through mutation, may not leave any mutation-free offspring (Butcher, 1995). The end result of either of these possibilities is that the population's least loaded class ceases to exist, leaving a slightly less fit class as its new least loaded class (Maynard Smith, 1988). After the zero mutation class is lost, the new least loaded class (that lineage having only one mutation) may suffer the same fate. Each time this happens it is analogous to the turning of a ratchet, because once the currently least loaded class is gone it can never be regenerated. As this ratcheting continues, the mutational load will gradually accumulate in the asexual population, until it becomes so great as to drive the population extinct. A sexual population, on the other hand, can always recreate its least loaded class through genetic recombination.

In contrast to the situation that one finds in a finite population, in an infinite population, *every* mutational class will be represented (including the zero class) according to a Poisson distribution with mean equal to U/s (Maynard Smith, 1988). This means that there will always be a large number of individuals carrying no mutations, and that the ratchet can not operate.

Muller's ratchet is an example of a group selection theory in which entire groups of asexual organisms will go extinct due to the irreversible accumulation of mutations.

Models of the ratchet which assume that mutations have multiplicative interactions There are many parameters which may affect the way in which the ratchet operates, or may prevent its operation altogether. The rate at which the ratchet turns (that is, the rate at which the population's least loaded line goes extinct due to random sampling error, and/or mutation) is dependent upon N, s, and U where N is the population size, "s" is the selection coefficient of the mutation when it is in a homozygous state, and "U" is the deleterious mutation rate per [haploid] genome (Bell, 1988a). In some models it also depends on the sign and degree of epistatic interactions between mutations (Charlesworth et al., 1993; Kondrashov, 1994c; Butcher, 1995) as well as population parameters such as carrying capacity and fecundity (Butcher, 1995).

It is worth mentioning at this point that there is considerable confusion in the literature as to how "U" actually is defined. In this review, U is defined as above. According to any number of reference texts, a genome is the entire *haploid* chromosome content of *one* complete set of chromosomes (Griffiths et al., 1999; Russell, 1998; Snustad et al., 1997). In spite of this fact, many authors use "U" interchangeable to mean either the genomic mutation rate or the total mutation rate, which, in diploids, would be 2U. This obviously leads to confusion because one can not unequivocally refer to the U value for an organism such as the yeast *Saccharomyces* which exists and reproduces both as a haploid as well as a diploid. Clearly, the U value for these two ploidy states can not be the same. In order to resolve this problem, we will use "U", as it has been defined by others, to refer to the [haploid] genomic mutation rate and if necessary $U_{(2n)}$ to refer to the total mutation rate of a diploid organism. According to this usage, $U_{(2n)} = 2U$.

To date, most models have assumed that deleterious mutations have multiplicative effects upon fitness. In multiplicative models, the ratchet clicks at a constant rate through time (Butcher, 1995). For instance, an individual having k mutations would have a fitness W(k) proportional to $(1 - s)^k$, where "s" is the selection coefficient against the mutation. Frequently, for convenience, the natural logarithm is used to express fitnesses such that $\ln W(k) = k \ln(1 - s)$. You will notice in these examples that each deleterious mutation has the same *initial* selective value. We will be coming back to this point later.

Felsenstein (1974) was the first to analyze the ratchet formally. He assumed that all mutations had equal effects, and that mutations interacted in a multiplicative manner. Using computer simulations, he found that deleterious mutations of very slight effect ($s = -0.001$) became fixed at a very high rate. However, since recombination had no effect upon the rate of fixation of these mutations, technically, this effect could not be credited to the ratchet per se. Instead, Felsenstein attributed this high rate to the random fixation of mutations through genetic drift in the virtual absence of selection. It should be pointed out that the fixation of each new mutation decreased the fitness of the population by a miniscule amount.

It turned out that the ratchet effect (i.e. the point at which recombination significantly affects the rate of incorporation of mutations) did not become apparent until the absolute value of the product of the effective population size and selection coefficient ($N_e s$) approached a value of approximately 1 (Felsenstein, 1974).

Felsenstein also found that mutations of a large effect (e.g. $s \approx -0.90$) resulted in the ratchet clicking at a very slow rate. Interestingly, it was the mutations of slight to intermediate effect (e.g. $s \approx -0.03$ to -0.10) which were the most detrimental to the fitness of the population. It was also these mutations of intermediate selective values which were most significantly affected by recombination. For $U = 1$, Felsenstein found that mutations having a selective value of $s \approx -0.10$ had the greatest contribution to the ill-health caused by the ratchet. This somewhat counterintuitive conclusion has been subsequently confirmed in computer simulations conducted by others (Butcher, 1995; Gabriel et al., 1993). Over a wide range of carrying capacities, and with a genomic mutation rate of $U = 1$, the mutations which contributed the most to the ill-health of the population were found to have selection values between $s \approx -0.05$ and $s \approx -0.30$. This somewhat counterintuitive finding is due to the fact that selection is ineffective against mutations having selective coefficients less than $1/N_e$ (where N_e is the effective population size) allowing such mutations to persist in a population and eventually become fixed. Mutations having a large (negative) effect on fitness are quickly eliminated by selection long before the line carrying them can become the least loaded line.

Haigh (1978) showed that in an infinite population, the distribution of the number of mutations per individual will follow a Poisson distribution. In finite populations, this distribution will only be approximated. It has been suggested that one of the most critical factors affecting the speed at which the ratchet turns is the number of individuals in the least loaded class (n_0), which can undergo stochastic variation from one generation to the next (Haigh, 1978). This number can be approximated by the following equation:

$$n_0 = Ne^{-U/s}$$

If N and s are small and U is large, then the ratchet will turn quickly. It will turn very slowly if U is small, or if N or s are large.

With this said, the next task is to determine whether the ratchet will proceed in natural populations, and if so, the speed at which it will occur. There are two methods by which this may be accomplished: 1) experimental estimates of the parameters such as U, and s (typically through mutational accumulation experiments in which lineages are propagated for numerous generations in the absence of recombination with minimal selection, and 2) inferential studies in which the operation of the ratchet can be deduced from comparisons between sister taxa one of which is subject to conditions promoting the ratchet and the other of which is not.

Experimental estimates of U and s We will begin by addressing first, the experimental estimates for U and s. Accurate estimates of these parameters are essential for numerous evolutionary and population genetics models (e.g. see Kondrashov, 1998), and yet, up until the 1990's, there was very little data on the rate of genomic deleterious mutation for any organism. For our purposes, these values are not only important for determining the potential effects of Muller's ratchet, but they are also critical in order to evaluated Kondrashov's Deterministic mutation model for the evolution of sex which is discussed in a later section. Over the past decade, a number of attempts have been made to estimate the genomic mutation rate of a variety of organisms. These studies have made it clear that genomic mutation parameters are neither easy to measure nor constant between taxa.

In order to place the current studies in perspective, it is of use to first review some of the earliest experiments which lent themselves to the estimation of U. These were the mutation accumulation experiments of Terumi Mukai, who employed a *Cy/Pm* balancer chromosome system of *Drosophila melanogaster* in order to maintain a wild type second chromosome sheltered from both selection as well as recombination in males heterozygous for this chromosome (Mukai, 1964; Mukai et al., 1972). Deleterious mutations were allowed to accumulate on these sheltered chromosomes for many generations. This second chromosome (which comprises about 40% of the euchromatic portion of the male *Drosophila* genome, Adams et al. (2000)) was subsequently exposed to selection in a homozygous state, and egg to adult viability was estimated. Assuming equal effects of mutations, the lower bound (U_{min}) for the mutation rate and the upper bound for the mean selection coefficient of mutations on that chromosome ($\bar{S}_{MAX}$) could be estimated from the per generation change in the mean and variance of fitness (Bateman, 1959). Extrapolation of Mukai's (1964) results yields a U_{min} of 0.35, with an average selection coefficient of $\bar{S}_{MAX} = 0.027$. Extrapolation of the results of a later study also yields a similar value of $U_{min} = 0.43$ with $\bar{S}_{MAX} = 0.023$ (Mukai et al., 1972).

Most subsequent *Drosophila* mutational estimates yielded lower values than those obtained by Mukai. Ohnishi (1977) found a U_{min} of 0.145 with an $\bar{S}_{MAX}$ of 0.030. A recent large scale *Drosophila* second chromosome mutation accumulation experiment yielded an estimate of the genomic deleterious mutation which was even lower, on the order of $U_{min} = 0.052$ with an average selection coefficients against these mutations estimated to be $\bar{S}_{MAX} = 0.113$ (Fry et al., 1999). Fry et al. suggest that their U_{min} estimates could be "a considerable underestimate of the overall rate of deleterious mutations" for at least two reasons. The first reason is due to the fact that that U_{min} is known to underestimate the true deleterious mutation rate if the assumption of equal effects of mutations is violated. In order to determine whether this assumption had been violated, they employed a technique first applied to this type of problem by Keightley (1994). Keightley used a maximum-likelihood approach to estimate deleterious mutation parameters such as U and s. This approach does not require the assumption of equal effects of mutations, and has been shown to produce more accurate results than the Bateman-Muikai method if the data do not conform to this assumption (Keightley, 1998). Both this approach as well as the minimum-distance approach of Garcia-Dorado (1997) seek to find the parameters of a theoretical distribution, such as a gamma distribution, which provide the best fit of that distribution to the observed data. Using a maximum-likelihood approach, Fry et al. (1999) were able to demonstrate that their data in fact, did not fit the expectations of an equal mutation model. Secondly, they went on to point out that neither their experiment, nor those of Mukai and Ohnishi measured the effects of mutations other than those that affect egg to adult viability in male flies, however this does not explain why their U_{min} estimate was so much lower than that of either Mukai or Ohnishi.

Starting in the latter half of the 1990's, researchers began to estimate genomic deleterious mutation rates in a variety of other organisms. In the nematode *Caenorhabditis elegans*, based upon changes in productivity following 60 generations of mutation accumulation, a very low estimate of $U \approx 0.0026$, with $s \approx 0.21$ was obtained using a maximum-likelihood approach (Keightley and Caballero, 1997). This estimate was later upwardly revised by the Keightley group to $U \approx 0.07$ (Davies et al., 1999). This revision was based upon their finding that in *C. elegans*, only about 4% of the deleterious mutations induced by EMS were detectable using the techniques employed in their previous experiment. The fact that greater than 96% of all deleterious mutations had a selection coefficient of less than 0.07% and were undetectable in mutation accumulation experiments implies that such experiments may consistently underestimate the actual genomic mutation rate. With this in mind, a 50 generation *C. elegans* mutation accumulation experiment, using a variety of life history traits to estimate the genomic mutation rate, placed U_{min} at 0.0225 with $s \approx 0.14$ (Vassilieva and Lynch, 1999), while a second, longer mutation accumulation experiment, lasting 214 generations yielded a U_{min} of 0.015 with $s \approx 0.12$ (Vassilieva and Lynch, 2000). Recently, the data from Keightley and Caballero (1997) and Vassilieva and Lynch (1999), have been reanalyzed using a maximum likelihood approach as well as the traditional Bateman-Mukai method (Keightley and Bataillon, 2000). Combining the two experiments, this reanalysis yielded an average U of approximately 0.005, with "s" varying from 0.13 to 0.51.

Deng and Lynch (1996) developed a method for the indirect estimation of population parameters such as U and s from changes in the mean and genetic variance of fitness traits of inbred lines. They subsequently used this technique to estimate U_{min} and $\overline{S}_{MAX}$, in the brine shrimp *Daphnia*, to be 0.74 and 0.14 respectively (Deng and Lynch, 1997).

Recently, the genomic mutation rate has been estimated for a diploid strain of the yeast *Saccharomyces cerevisiae.* Maximum-likelihood estimates placed U at approximately 9.5×10^{-5} with "sh" estimated to be 0.217 ("sh" is the effect of mutations in a heterozygous state) (Zeyl and De Visser, submitted manuscript).

In prokaryotes, U_{min} has been estimated through mutation accumulation experiments to be as low as 0.00017 in *E. coli* (Kibota and Lynch, 1996). The nearly 1500 fold difference between *E. coli* and *Drosophila* genomic mutation rates is likely due in large part to the 50 fold difference in haploid genome size and the fact that there are 25 times as many cell divisions per generation in *Drosophila* are there are in *E. coli* (Kibota and Lynch, 1996).

There have been relatively few estimates of U made for plants, however one study estimated that U_{min} was on the order of approximately 0.05 in *Arabidopsis*, with an $\bar{S}_{MAX}$ of 0.20 (Schultz et al., 1999) A second study failed to detect any significant difference between control and mutation accumulation lines (Shaw et al., 2000).

In Primates, inferences based upon the rate of occurrence of hemophilia B have led to a surprisingly high estimate that U may be as large as 2.0 in humans (Giannelli et al., 1999; Giannelli and Green, 2000). This incredibly high estimate of U is given support by another study involving the direct sequencing of pseudogenes, which yielded estimates of $U \approx 1.5$ in humans (Nachman and Crowell, 2000). Molecular divergence data has yielded a somewhat lower estimate of $U \approx 0.8$ in humans, $U \approx 0.85$ in chimpanzees, and $U \approx 0.60$ in gorillas (Eyre-Walker and Keightley, 1999). Such high rates of mutation in both humans and the great apes suggests that these mutations may interact synergistically rather that multiplicatively, leading to a "quasi-truncation selection", which, mediated by recombination, allows the selective removal of a larger number of deleterious mutations with relatively fewer genetic deaths than would be possible in an asexual population (Kondrashov, 1982; Kondrashov, 1994c; Eyre-Walker and Keightley, 1999; Giannelli and Green, 2000). If such were not the case, slowly reproducing species, as exemplified by the great apes, would be driven to extinction by their mutational loads (Eyre-Walker and Keightley, 1999). For a contrasting opinion on the importance of synergistic epistasis on the rate at which Muller's ratchet proceeds in slowly reproducing species see Butcher (1995), as discussed below.

In concluding this section on the estimation of genomic mutation rates, it is important to mention that the some what confusing state of this field is not aided by the lack of standardization of nomenclature nor by the lack of similarity between experimental approaches to the measurement of U. It is not at all clear that the values obtained for U in *Drosophila* (which are based upon egg to adult viability) are comparable to values obtained in other organisms, such as *C. elegans*, using measures such as "productivity". Many of the studies cited have used the Bateman-Mukai method to estimate the deleterious genomic mutation rate. To the extent that deleterious mutations fail to abide by the assumption of "equal effects", these studies will underestimate U and over estimate "s" (Keightley, 1998). Even if we ignored all of the aforementioned problems associated with making estimates regarding the deleterious genomic mutation rates, a study by Shabalina et al. (1997) clearly demonstrates another problem; namely the fact that mutations that are benign in a benign environment, may very well be deleterious in a harsher, more realistic environment. If this is the case, then what are we to infer from the numerous mutation accumulation experiments carried out in rather benign laboratory environments?

For those interested in recent reviews on the measurement of spontaneous mutation rates see Kondrashov (1998), Keightley and Eyre-Walker (1999), Rosche and Foster (2000) and Bataillon (2000). For a review of a number of studies attempting to estimate genomic deleterious mutation rates see Keightley and Eyre-Walker (1999). For an early review on the effects of mutation on fitness in *Drosophila* see Simmons and Crow (1977), and for a recent survey of mutation rates per genome, per genome replication (μ_g) across a wide variety of taxa, see Drake (1999).

Table 5 below summarizes the genomic mutation estimates that can be drawn from a number of studies.

Using Mukai's estimates (Mukai, 1964; Mukai et al., 1972) of the rates of mutation and the effect of viability modifiers in *D. melanogaster*, Bell (1988a) concluded that the most reasonable values for U/s were somewhere between 10 and 20. Using this as a starting point, Bell calculated the rate at which the ratchet would turn given different population sizes (N), and for various values of U/s. His conclusion was that an asexual population of 10^{10} organisms would be driven to extinction by the ratchet within approximately $10^{6.5}$ generations. Organisms having population sizes smaller that 10^{10} individuals will suffer the effects of the ratchet even more quickly. Since organisms having a body weight of one gram or greater are not likely to have populations as large as 10^{10}, Bell concludes that the evolution of sex must have occurred prior to the evolution of large body size.

Large population size is not the only way in which a population can evade the ratchet. If the mutation rate per genome, U, is reduced, this too will slow the ratchet. Since there is a lower limit to the fidelity achievable during DNA replication (Bell, 1988a; Bernstein et al., 1981), the only means of achieving a further reduction in the per-genome mutation rate is to reduce the size of the genome itself. While such a reduction in genome size will not reduce the rate of mutation at any particular locus, it will reduce the overall genomic mutation rate simply because there will be fewer

TABLE 5. The Genomic Mutation Rate $U_{(n)}$ with Selection Coefficient "s" where Applicable

Species	Experimental Method	$U_{(n)}$		s	Reference
Drosophila	Mutation Accumulation	0.35	BM	0.027	Mukai (1964)
Drosophila	Mutation Accumulation	0.43	BM	0.023	Mukai et al. (1972)
Drosophila	Mutation Accumulation	0.145	BM	0.030	Ohnishi (1977)
Drosophila	Mutation Accumulation	0.052	BM	0.113	Fry et al. (1999)
Drosophila	Molecular Divergence	0.028		N.A.	Keightley & Eyre-Walker (1999)
Saccharomyces	Mutation Accumulation	9.5×10^{-5}	ML	0.217*	Zeyl & De Visser (Unpublished)
E. coli	Mutation Accumulation	0.00017	BM	0.012	Kibota & Lynch (1996)
C. elegans	Mutation Accumulation	0.0026	ML	0.10	Keightley & Caballero (1997)
C. elegans	Mutation Accumulation	0.0225	BM	0.14	Vassilieva & Lynch (1999)
C. elegans	Re-analysis	0.07		0.14	Davies et al. (1999)
C. elegans	Mutation Accumulation	0.015	BM	0.12	Vassilieva et al. (2000)
C. elegans	Re-analysis	0.005	ML		Keightley & Bataillon (2000)
Arabidopsis	Mutation Accumulation	0.05	BM	0.20	Schultz et al. (1999)
Arabidopsis	Mutation Accumulation	0		N.A.	Shaw et al. (2000)
Daphnia	Inbreeding Experiment	0.74		0.14	Deng & Lynch (1997)
Homo sapiens	Molecular Sequence Data	1.5		N.A.	Nachman& Crowell (2000)
Homo sapiens	Molecular Divergence	0.8		N.A.	Eyre-Walker & Keightley (1999)
Homo sapiens	Hemophilia based Inference	2.0		N.A.	Giannelli & Green (2000) Giannelli et al. (1999)
Pan troglodytes	Molecular Divergence	0.85		N.A.	Eyre-Walker & Keightley (1999)
Ribo viruses	Other	0.76**		N.A.	Drake & Holland (2000)
Gorilla gorilla	Molecular Divergence	0.60		N.A.	Eyre-Walker & Keightley (1999)

* Represents the average heterozygous selection coefficient.
** Represents μ_g the deleterious mutation rate per genome replication.
BM = Bateman-Mukai method.
ML = Maximum Likelihood method.

nucleotides to mutate. It has been suggested that this may explain why the mitochondria of metazoans have evolved such small genome sizes (Bell, 1988a; Maynard Smith, 1988). Bell points out that while human cells contain a closed population of about 10^3 mitochondria, (potentially an ideal situation for the operation of the ratchet), the mitochondrial genome is only a tiny fraction the size of the nuclear genome. So, while U/s for the nuclear genome may take on a value somewhere between 10 and 20 (for s = 0.01), a comparable value of U/s for the mitochondrial genome may be closer to 0.0015. Most of the population would consist of individuals in the least loaded mutation class, and the ratchet would turn very slowly. Interestingly, it appears that the mitochondrial mutation rate in *C. elegans*, is fairly high, on the order of 1.6×10^{-7} per base-pair per generation (Denver et al., 2000), as compared to the nuclear mutation rate of 2.3×10^{-10} per base pair per genome replication (Drake, 1999). In the face of this rather high a mutation rate, if it weren't for the fact that mitochondrial genome of *C. elegans* consists of less than 14k.b. (Okimoto et al., 1992), it would likely be quite vulnerable to the ratchet.

While metazoan mitochondria may be able to evade the ratchet because of their small stripped-down genome sizes, this may not apply to yeast mitochondria. Not only do yeast mitochondria occur in very small populations (approximately 20 per cell), but their genomes are large. At 86 kb, they are about five times the size of metazoan mitochondria. As Bell (1988a) put it, "The proposition that small genomes evade the ratchet can be turned on its head: because of the ratchet, large genomes cannot evolve in the absence of recombination." Perhaps evasion of the ratchet is the reason the mitochondria of yeast recombine at a high rate (Birky, 1978; Rowlands and Turner, 1974). Relatively high rates of recombination also occur in the large chloroplast genomes of *Chlamydomonas reinhardti* (Harris, 1989). It will be instructive to find whether other large mitochondrial genomes such as those of the flagellate *Reclinomonas americana* (69k.b.) (Lang et al., 1997) or the fungus *Podospora anserina* (100k.b.) (Cummings et al., 1985) also undergo recombination.

Heller and Maynard Smith (1979) have found that the ratchet can operate even in a diploid selfing population. In this case, the diploid selfing population behaves as would an asexual haploid population. This is because each mutation in the selfing population will become fixed or lost with an equal probability. Maynard Smith (1988) points out that those mutations that do become fixed are analogous to the mutations that occur in an asexual haploid.

Models of the ratchet involving synergistic epistasis between mutations Virtually all analyses of the ratchet have assumed that there is no interaction between mutations, that is, that the effects of multiple mutations are independent. In other words, most models assume that there is no epistasis (epistasis is defined here as any departure from multiplicative fitness interactions between mutations on a normal scale, or from additive interactions between mutations on a log scale). The assumption of independence may or may not be valid (see studies cited in the section on the Deterministic Mutation hypothesis). For this reason, a number of researchers have analyzed the effects that mutations, interacting in a synergistic manner, have on Muller's ratchet. Kondrashov (1994) analyzed such a situation, concluding that even moderate synergistic epistasis was sufficient to significantly slow the ratchet, and strong synergism could effectively halt the ratchet. He explained this by noting that synergism between mutations would tend to eliminate those clones having the greatest number of mutations. This would effectively increase the proportion of the population consisting of the least loaded classes of individuals. Since the rate at which the ratchet turns depends critically upon the size of this least loaded class, the larger this class, the less likely it is to be lost due to "sampling error".

Butcher (1995) had a slightly different way of explaining Kondrashov's results. He pointed out that after a certain number of mutations have accumulated in an asexual lineage's least loaded line, a time will eventually be reached at which, due to synergism between mutations, the effect of the next mutation is too large to drive the ratchet because it is immediately eliminated. At this point, the rate of fitness loss of the population due to the ratchet becomes zero. While the population's fitness may be low, it may nevertheless persist indefinitely.

Charlesworth et al. (1993) came to conclusions similar to those of Kondrashov after running computer simulations involving both multiplicative as well as synergistic interactions between deleterious mutations. They too found that the ratchet was "clearly slowed" by synergism. Interestingly, while the ratchet clicks at a constant rate in the multiplicative model, its rate actually slows down as the population's fitness decreases in models which assume synergism between mutations (Butcher, 1995).

Butcher (1995) has also investigated the effects of synergistic epistasis on the ratchet, using computer simulations. Previous models (Kondrachov, 1994c; Charlesworth et al., 1993) had assumed that all mutations have the same *inherent effect* (i.e. their effect prior to modification through epistatic interactions with other mutations). Butcher, on the other hand, made the reasonable assumption that mutations would exhibit a continuous distribution of inherent effects, ranging from almost neutral to lethal. Butcher concluded from his simulations that . . .

> "...epistasis will not halt the ratchet provided that rather than a single deleterious mutation effect, there is a distribution of deleterious mutation effects with sufficient density near zero."

In other words, synergtistic epistasis could not alleviate the effects of the ratchet, provided that there were many mutations of very slight effect. Synergism would enhance the effect of these mutations to the point at which they could drive the ratchet.

To help explain these conclusions, Butcher begins by reminding us that in the standard models of Muller's ratchet (i.e. those in which the effects of mutations are assumed to interact in a multiplicative manner), mutations having a very slight deleterious effect will be fixed at quite a high rate. While the rate of incorporation of mutations into the population may be high, the actual effects on the populations fitness are negligible. On the other hand, mutations which are much more deleterious are selectively removed much more efficiently. While such mutations may occasionally result in a click of the ratchet (decreasing the population's fitness considerably in the process), the frequency of such clicks is "vanishingly lower" (Butcher, 1995). It is therefore (as we have already noted) mutations of "intermediate, slightly deleterious effect" which contribute the most to the ratchet. These facts are important in explaining the results of both Kondrashov (1994c) and Charlesworth et al. (1993) with respect to the apparent ability to halt the ratchet through synergistic epistasis. Butcher explains this by noting that, if one assumes that all mutations have the same inherent effect, as an asexual population's least loaded line gradually accumulates more and more mutations, the effect of synergism between these mutations will be to increase their *actual effect* to the point at which they will no longer be able to drive the ratchet because they are selectively removed. On the other hand, if mutations have a continuous distribution of selection coefficients, synergism will increase the actual effects of all mutations. The actual effect of mutations at the high side of such a distribution will be increased to a such a degree that they can no longer be able to drive the ratchet. These mutations will be selectively removed. At the same time however, synergism will also increase the actual effects of mutations which previously had nearly negligible effects. It is these mutations that will now be able to drive the ratchet.

Butcher (1995) also found a very interesting relationship between the population's growth parameter R and the mean time to extinction. He defined R as the maximal rate of increase of a population in the absence of any mutations. It is equivalent to a measure of initial maximum fitness. In the absence of epistasis, the mean time to extinction is a linear function of ln(R). With epistasis however, the mean time to extinction increases exponentially as ln(R) increases. Butcher explains this by noting that an increase in R results in a greater skew of the distribution of actual effects at low maximum fitness (that is when the most fit class has a low fitness). With synergistic epistasis, any distribution of inherent mutation effects can be skewed by an R sufficiently large that the number of slightly deleterious mutations is in equilibrium with the number of newly arising back mutations and suppressor mutations. Another way of looking at this relationship is to note that in asexual organisms of low fecundity, epistasis should have little effect upon the mean time to extinction. On the flip side of the coin, high fecundity can actually greatly enhance the longevity of a population given that some synergistic epistasis is present.

This is an important finding. It suggests that asexual populations can evade the ratchet by reproducing at a very high rate (which, perhaps not surprisingly, is what many asexual organisms do). Butcher is quick to point out, however, that R is not determined solely by the rate at which offspring are produced, but rather by the rate of production of offspring that survive to compete and reproduce.

Models of the ratchet involving antagonistic epistasis between mutations
Wagner and Gabriel (1990) modeled a different type of interaction between mutations. They begin by contrasting conditional and unconditional mutations. An unconditional mutation is one which cannot be compensated for by other mutations. In other words, the original phenotype cannot be restored by a suppressor mutation at another locus. Conditional mutations are mutations, the effects of which, can be compensated for by suppressor mutations. Wagner and Gabriel's model addresses only the conditional mutations that contribute to Muller's ratchet. The model assumes that while deleterious mutations accumulate as a result of the ratchet and drift, their effects can be ameliorated through the accumulation of beneficial compensatory mutations. Compensatory mutations, they argue, are much more frequent than back mutations, which are indeed assumed to be very rare. Deleterious mutations are assumed to have multiplicative effects upon fitness. As more and more deleterious mutations accumulate, the probability of compensatory mutations increases. Simulations showed that under certain conditions compensatory mutations were as effective as recombination in combating the effects of the ratchet. While Wagner and Gabriel concluded that in the short run compensatory mutations may prevent an asexual population from being driven extinct by the ratchet, they conceded that in the long run it is unclear whether such a population could become "saturated" with compensatory mutations, and no longer be able to evade the downward spiral in fitness caused by the ratchet. It should also be noted that by definition, no amount of compensatory mutation can ameliorate the

effects of *unconditional* mutations accumulated in a non-recombining population through the ratchet.

Muller's ratchet and the evolution of the sex chromosomes (diploid dioecy) Muller was the first to propose an explanation for the evolution of sex chromosome dimorphism, and the inertness of the Y chromosome (Muller, 1914; Muller, 1918; Muller, 1932). According to Muller (1918), both the X and Y chromosomes were initially alike. At some point in time, however, a genetic change occurred that caused a reduction in recombination between these chromosomes. A balanced lethal could result in selection pressure for such a reduction in recombination, since recombinants would be lethal, and thus only non-recombinant X-Y karyotypes would survive. From this point on, the genes on the incipient Y, as they acquired further mutations, would in effect always be in a heterozygous state, and therefore any deleterious recessive mutations occurring within these genes would be sheltered from selection. The accumulation of such mutations would eventually lead to widespread inactivation (or "inertness") and degeneration of the Y chromosome (Muller, 1918). The X chromosome, meanwhile, would be spared from a similar accumulation of recessive mutations because it would be continually subjected to selection in the heterogametic sex. As the two sex chromosomes diverged in sequence, their synaptic attraction would diminish.

Fisher (1935) noted that the proto-X chromosome in Muller's scenario would be subject to the same rate of deleterious mutation as the proto-Y chromosome. Deleterious recessive mutations on the Y chromosome would be selectively removed whenever a similar mutation occurs in the same locus on the X chromosome. He also showed that at equilibrium, in an infinite population, the frequency of recessive lethals would be the same for the two chromosomes. Nei (1970) re-analyzed the same situation, and concluded that in finite populations it is possible for deleterious recessive mutations to accumulate more rapidly on the Y chromosome after all. This is due to the fact that in small populations ($N_e < \sim 4{,}000$), most recessive lethals will be eliminated from the X chromosome by a combination of selection and genetic drift. Thus, given a small enough effective population size, the Y chromosome would indeed be sheltered from selection, and deleterious recessive mutations could accumulate.

Charlesworth (1978) argued against the hypothesis proposed by Muller and supported by Nei for the evolution of sex chromosomes. He felt that it was not plausible, because the effective population size of most species is thought to be much larger than that required to shelter the Y according to Nei's calculations. Instead he preferred to invoke Muller's ratchet to explain the degeneration of the Y chromosome.

According to his model, the population of Y chromosomes would gradually lose their least loaded line due to random sampling. This would represent one turn of the ratchet. The frequency of deleterious mutations per locus on the Y chromosome would therefore gradually increase. This increase in itself, however, is not responsible for the degeneration of the Y. Instead, Charlesworth invokes dosage compensation. At some point in time, the population of Y chromosomes contains enough "impaired" loci to create a selective advantage for any gene that is able to boost the level of the gene products of undamaged X chromosome loci, thereby compensating for the defective gene products produced by the homologous Y loci. Once the X chromosome's gene products had been so enhanced, there would be further selection to reduce the level of expression of the Y loci, thereby restoring the proper balance between the sex chromosomes and the autosomes. Once the quantities of gene products of Y chromosome loci has been sufficiently reduced, selection would no longer prevent the genes located on that chromosome from degenerating. The end result would be an inert Y chromosome.

Experimental evidence for Muller's ratchet Evidence in support of Muller's ratchet is gradually accumulating. Bell (1988a) cites a large body of work on the senescence of ciliated protozoa in the absence of sexual conjugation as evidence for the occurrence of Muller's ratchet in these organisms. Chao (1990) has shown that Muller's ratchet can occur when segmented RNA bacteriophages are passed through genetic bottlenecks. Twenty lineages of the three-segmented RNA bacteriophage $\phi 6$ were passed through a series of bottlenecks followed by growth cycles. Each growth cycle was initiated with a single individual, which was then allowed to grow up to approximately 8×10^9 individuals. At the end of forty growth cycles, representatives of each of the transferred clones were competed against their genetically marked parental clones in order to measure any fitness changes which might have occurred. Chao found that the mean relative fitness of the transferred clones declined significantly to 78% of that of their parental clones. He pointed out that while dispersal has often been invoked as an impetus for the evolution of sex (the generation of genetic variation aids the ability of organisms to spread through a heterogeneous environment), Muller's ratchet may play a significant role in the evolution of organisms such as viruses in which dispersal often results in a severe bottleneck. Recombination can occur in segmented viruses such as $\phi 6$ whenever two or more viruses co-infect a single cell. Chao's protocol effectively eliminated the possibility of recombination, and as his results show, Muller's ratchet operates rather quickly under such circumstances.

In a subsequent experiment, Chao et al. (1992) found that, as predicted, recombination was able to enhance (although only modestly) the fitness of $\phi 6$ viruses which had earlier undergone fitness losses due to the ratchet.

Duarte et al. (1992) performed experiments similar to those performed by Chao, using the non-segmented vesicular stomatitis virus (VSV), and obtained similar results in only twenty transfers. While sharp decreases in fitness were seen when VSV was put through genetic bottleneck transfers, the opposite effect was observed when large numbers of viruses were transferred (Holland et al., 1991). When approximately10^4 virus particles were transferred at each passage, VSV showed rapid fitness increases. Duarte et al. (1992) suggest, as did Chao (1990), that Muller's ratchet may play a significant role in the evolution of highly mutable RNA viruses. In their view, the ratchet may help extremely pathogenic viruses to survive, by conferring the ability to "ratchet down fitness (and virulence) during occasional bottleneck transmissions," thereby ensuring the survival of at least some of their hosts.

More recently, Andersson and Hughes (1996) have investigated whether Muller's ratchet can proceed in a prokaryotic organism (*Salmonella typhimurium*) having a much lower mutation rate that the RNA viruses examined by Chao (1990), Chao et al. (1992), Holland et al. (1991), and Duarte et al. (1992). Andersson and Hughes put 444 lineages through an average of 60 single-cell bottlenecks. They found approximately one percent of these lineages suffered a significant loss in fitness as measured by growth rate. They calculated the mean spontaneous mutation rate to be 0.3 to 1.5×10^{-9} per base pair per generation.

A number of different animal lineages harbor a variety of endosymbiotic bacteria which are passed on maternally through cytoplasmic inheritance (Moran, 1996). The concordance between the phyogenies of these endosymbionts and their hosts suggest that these associations are ancient. The primarily parthenogenetic group of plant sucking insects known as aphids harbor several different types of endosymbiotic bacteria. One of these, *Buchnera aphidicola*, an obligate endosymbiont, thought to have been passed on exclusively through maternal cytoplasmic inheritance with little if any recombination for some 100 to 250 million years (Moran, 1996).

Typically, endosymbiotic bacteria have small effective population sizes, pass through a bottle-neck each generation, and have little if any recombination; a potentially ideal situation for the operation of the ratchet (Moran, 1996). To investigate whether the ratchet was operating in these endosymbiotic prokaryotes, Moran (1996) compared the rate of substitution of 16S rDNA of five endosymbiotic lineages to the rate of substitution in their free-living counterparts. In each case, the endosymbiont's 16S rDNA evolved significantly more rapidly that its free-living counterpart.

Synonymous and non-synonymous substitution patterns were also compared between two *Buchnera* species and their free-living counterparts *E.coli* and *Salmonella typhimurium*. The synonymous substitution rates between the two *Buchnera* species were similar to those between *E.coli* and *Salmonella typhimurium*, a finding which is in agreement with the presumption that the divergence times between the two *Buchnera* and the two free-living enterics are similar. However, when non-synonymous substitution rates were compared, the endosymbiotic *Buchnera* had consistently higher rates than their free-living counterparts in all five *trp* genes examined. This finding supports the idea that the increased rate of evolution in these endosymbionts is due to an accumulation of amino acid changing (deleterious) substitutions driven by Muller's ratchet (Moran, 1996).

In another study, it was found that the stabilities (as determined by $-\Delta G$) of domain I of the 16S rRNA subunits were 15 to 25% lower in six different endosymbiont lineages that in their free-living counterparts (Lambert and Moran, 1998). This too supports the hypothesis that the ratchet is operating in these small, non-recombining populations of endosymbionts, resulting in the accumulation of mildly deleterious mutations, even in the highly conserved 16S rDNA sequence.

Shabalina et al. (1997) found evidence for the operation of the ratchet in two panmictic populations of *D. melanogaster* which were maintained under "relaxed" selection. One of these populations was kept under benign conditions while the second was kept under harsher conditions. After 30 generations, the number of surviving offspring per female declined by 0.2% per generation in the population kept under benign conditions, while it had declined by 2.0% in the population kept under harsher conditions. These results are important in that they demonstrate the differential effects of mutations under harsh and benign conditions. It seems likely that in nature, very few mutations would find themselves in a benign environment. This suggests that researchers wishing to estimate U through mutation accumulation experiments, may, in the future want to measure fitness effects in a harsh (i.e. realistic) environment rather than in the typically benign conditions found in the laboratory.

Mitotic recombination and the ratchet In some completely asexual organisms, mitotic recombination may play a role in ameliorating the effects of Muller's ratchet. It has been suggested that biased gene conversion evolved in order to reduce the genetic load resulting from certain common types of mutation (Bengtsson, 1990, 1985), however to date it is unclear as to how such a conversion mechanism would be able to "distinguish" the mutant allele from the wildtype allele and therefore be able to bias conversion in favor of the wildtype allele. Nevertheless, mitotic recombination may be viewed as a sort of pseudo-sex in that an asexually reproducing diploid organism can readily produce progeny containing fewer mutations than it itself simply through the random conversion of a locus heterozygous for a mutation to a locus homozygous for the wildtype. As opposed to the conventional view, which regards the ratchet as operating in virtually all asexual lineages, it may be more productive to view it as operating in a variety of genetic situations that lie somewhere along a continuum of varying degrees of sexuality. It has been found that in petite (respiration incompetent) yeast reproducing mitotically, the ratchet appears to operate, while it does not in grande strains reproducing in a similar fashion (Selk and Wills, 1997). It will be informative to determine whether different levels of mitotic recombination in the petite and grande strains contribute to this phenomenon. One possibility is that the higher metabolic levels of grande cells support higher levels of the enzymes involved in mitotic recombination and various repair systems.

Other mutational processes Muller's ratchet is not the only way by which a population's mean fitness can be decreased through the accumulation of mutations. Charlesworth et al. (1993) point out the importance of distinguishing between the *loss* of the class with the fewest mutations (the ratchet) and the *fixation* of mutant alleles that occurs in small populations. The distinction is that with the ratchet, one can have a population with a very large number of accumulated mutations, in which no single mutation is found at high frequency. This is in contrast to the deleterious effect caused by the fixation of mutations in small populations. According to Charlesworth et al. (1993), "fixation is greatly slowed by increased population size, whereas the ratchet process is more strongly dependent on restricted recombination but less sensitive to population size."

Another process which can lead to the degradation of the fitness of a genome occurs when a beneficial mutation arises and spreads to fixation dragging along with it any deleterious mutations associated with the genome in which it first appeared (Lynch and Blanchard, 1998). Since the likelihood of such a mutation arising in a lineage's least loaded line is small, such a mutation will carry to fixation all of the deleterious mutations that were present in genome in which the mutation first appeared. In other words, the deleterious mutations will be able to "hitch-hike" to fixation along with the beneficial mutation. Lynch and Blanchard (1998) point out that if the selective environment subsequently changes, such that the once beneficial mutation is no longer advantageous, then the lineage in which the mutation had become fixed could suffer a decline in fitness.

The "Mutational Meltdown" Some authors have pointed out problems with the current models that invoke mutation as the driving force behind the maintenance of sex (Lynch and Gabriel, 1990; Melzer and Koeslag, 1991; Lynch et al., 1993; Gabriel et al., 1993). Lynch and Gabriel (1990) point out that all contemporary models involving mutational load have one rather strange thing in common:

> "...despite their deleterious nature, the mutations have no influence on the demographic features of the population that harbor them. Selection acts on the relative contributions of individuals to the next generation, but population-level fertility is assumed to be effectively infinite. The number of reproducing adults is fixed, and the population is restored to its carrying capacity each generation. There is no extinction."

According to Lynch et al. (1993; Lynch and Gabriel, 1990), a more realistic model of Muller's ratchet would take into account the effect of an increasing mutational load on population dynamics. They make the reasonable assumption that as the mutational load builds up in an asexual population, there would be a gradual decrease in population size. This would increase the effects of random genetic drift, in particular the fixation of more and more deleterious mutations. The end result will be the extinction of the population due to an accumulation of an unsupportable genetic load. This process they dubbed a "mutational meltdown" (Lynch et al., 1993; Lynch and Gabriel, 1990). They go on to state that ...

> "... there can be no equilibrium mutation load or population size unless the time to extinction is great enough to allow a substantial likelihood of back-mutation." (Lynch and Gabriel, 1990).

In conventional models of Muller's ratchet, for a fixed μ the ratchet is more damaging when s is small. In the mutational meltdown model, in contrast, large values of s are more damaging. This is because, although the ratchet turns more slowly with a large s, this is more than offset by the proportionately worse survival of individuals with each turn of the ratchet. This leads to a reduced population size, and the consequent increased likelihood of fixation of deleterious mutations.

The Deterministic Mutation Model In the early 1980's, Alex Kondrashov developed a new hypothesis for the maintenance of sexual recombination, the Deterministic Mutation model (Kondrashov, 1982; Kondrashov, 1988). The important distinction between Muller's ratchet and the ostensibly similar *deterministic mutation hypothesis* is that Muller's ratchet relies upon finite populations which are at least occasionally of small enough size that the least loaded lineage may be lost simply through sampling error. The deterministic mutation hypothesis does not require such size bottlenecks.

It is based instead upon an infinite population model in which the total deleterious mutation rate $U_{(2n)}$ is greater than 1 for diploids (Keightley and Eyre-walker, 2000). It assumes that there is synergistic epistasis between mildly deleterious mutations, such that the negative effect of two mutations is greater than the sum of their individual effects. (You will recall that epistasis is defined as any departure from multiplicative fitness interactions between mutations on a normal scale, or additive interactions between mutations on a log scale). This model also assumes that truncation selection is operating on individuals having a threshold number T or more mutations. Under such conditions, the advantage of sex can be sufficiently great to overcome the twofold cost that is sometimes associated with it. In this model, the advantage of sex increases with the rate of mutation, and with decreasing effect of individual mutations. Kondrashov (1982) explains this by noting that in a sexual population containing a large number of mutations (each of which is only slightly deleterious by itself), the death of each individual having T or more mutations removes a large number of mutations from the population. This increases the sexual population's fitness relative to that of an asexual population because recombination is constantly regenerating the variance in the number of mutations per individual following threshold selection. This effect can be visualized in Fig. 7.

This figure (modified from Maynard Smith (1988)) gives an example of the manner in which the deterministic mutation model may function. It shows the frequencies of individuals carrying different numbers of deleterious mutations *before* selection. T is the threshold for truncation selection. Individuals carrying T or fewer mutations suffer no ill effects, while individual carrying more than T mutations die. As can be seen, the sexual population will be able to purge itself of many more mutations per "genetic death" than the asexual population.

Is there any evidence either for or against the Deterministic Mutation hypothesis? Unfortunately, as we saw earlier, estimates of the per genome mutation rate U are scattered over a wide range of values (see table 5). Some organisms, such as the great apes, clearly have a total genomic mutation rate $U_{(2n)}$ greater than 1, and are therefore likely candidates for the operation of deterministic mutation. However, it is not clear that deterministic mutation alone is maintaining sex in this group. In this group, sex may be maintained simply the fact that they can't reproduce without sex (due to genomic imprinting for example)—a constraint of their evolutionary history.

Other groups of organisms, such as *C. elegans*, clearly have genomic mutation rates of less than 1, and sex in these organisms is therefore unlikely to be maintained by deterministic mutation.

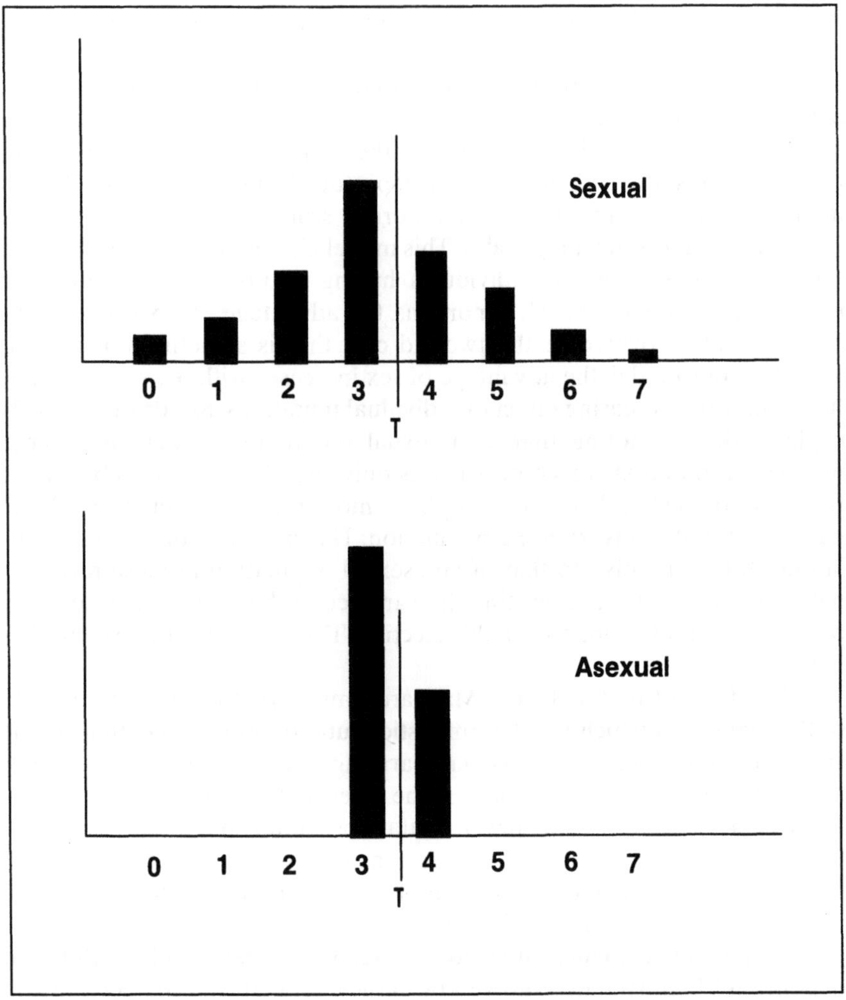

FIG. 7. Variation in the frequency of individuals carrying different numbers of deleterious mutations in sexual and asexual populations before threshold selection removes all individuals having four or more mutations.

Because of the overwhelming importance of the parameter U to a variety of the mutational hypotheses regarding the evolution of sex, it is clear that more research and data are required before the validity of these hypotheses (or lack thereof) can be determined. For a lively debate on the current status of the Deterministic Mutation hypothesis see Kondrashov (2001) and Keightley and Eyre-Walker (2001).

Another critical assumption of the Deterministic Mutation hypothesis is that deleterious mutations act synergistically with one another. Is there evidence to support this assumption? There is some, although this evidence too is somewhat unclear. De Visser et al. (1996) showed that synergistic epistasis was occurring between pre-existing and newly induced mutations in the haploid, facultatively sexual algae *Chlamydomonas moewusii*. However, this conclusion relied upon a number of assumptions that could not be directly verified. A subsequent series of experiments by De Visser et al. (1997a) found evidence suggesting that deleterious mutations affecting the carrying capacity (K) were synergistic in *C. moewussii*. They could not, however, demonstrate synergism between mutations affecting the intrinsic rate of increase (r). They speculated that their findings might shed some light on the observation that sexual reproduction is frequently found in constant environments in which K-selection is prevalent, while asexual reproduction is more frequently found in variable environments in which r-selection is more prevalent.

Clark and Wang (1997) examined the genetic interactions of P-element-induced mutations in *Drosophila melanogaster*, and found significant epistatic interactions in 35 of 128 or 27% of the comparisons. It was not clear from their paper, however, what proportion of these interactions were synergistic, and what proportion were antagonistic.

Mukai (1969) found evidence of synergistic epistasis between mutant polygenes affecting viability that were located on chromosome 2 of *D. melanogaster*. It should be noted that these polygenes had been made homozygous by balancer chromosome crosses in this set of experiments. Mukai pointed out, however, that while there was no direct evidence for synergistic epistasis in the heterozygous state, the data from some 1.5 million flies that had been presented in an earlier paper (Mukai and Yamazaki, 1964) was wholly consistent with this possibility.

In a study of the effects of inbreeding in the Monkey Flower, *Mimilus guttatus*, Willis (1993) found some evidence of synergistic epistasis in inbred lines. Offspring were scored for several fitness components: germination, flowering, flower production, and pollen viability. The natural logarithms of these fitness components were then plotted for varying degrees of inbreeding ($F = 0$, $F = 0.25$, $F = 0.50$, and $F = 0.75$). If there were independent mutation effects, there would be a negative linear relationship between these two variables. This is exactly what Willis found for three of the fitness measures: germination, flowering, flower production. For pollen viability, however, there was evidence of significant synergistic epistasis.

Elena and Lenski (1997) recently performed an elegant set of experiments to test for synergism among deleterious mutations in *E. coli*. They used a modified transposon, Tn10, to generate new, essentially random insertional mutations. They created a total of 225 different mutant strains. One third of these had one mutation, one third had two mutations, and the remaining third had three mutations. Double and triple mutants were created by repeated rounds of transposon-induced mutagenesis. Relative fitnesses were measured by performing three competitions between each mutant strain and its non-mutated progenitor strain. They found a log linear decrease in fitness with increasing number of mutations. This implies that these mutations interact with each other to affect fitness in an additive (not multiplicative) manner on a log scale. This is to be expected if the effects of multiple mutations are independent of each other. In fact, Elena and Lenski's results showed no evidence of interactions between mutations. This, they pointed out, could be due to three possibilities: 1) there were in fact no interactions between mutations, 2) the design had insufficient statistical power to detect any interactions, or 3) there were interactions, but the number of antagonistic interactions was approximately equal to the number of synergistic interactions. On average then, the effects of these mutations would tend to cancel each other out, producing a log linear fitness curve.

What does all of this mean for the Deteministic Mutation hypothesis? Otto and Feldman (1997) have demonstrated mathematically that if loci vary in the sign and strength of their interactions with each other, this reduces the likelihood of recombination evolving as a response to mutation. Elena and Lenski (1997) designed a second set of experiments in order to determine the relative frequency of synergistic versus antagonistic epistasis between deleterious mutations. In this series of experiments, double mutants were created not by repeated mutagenesis, but by using transduction to create various pairs of mutations, each of which was derived from single mutant strains. In this way, the interaction of pairs of mutations could be compared to their expected effects assuming no epistasis. It was found that in three out of the 27 comparisons there was significant synergistic epistasis, and in another four there was significant antagonistic epistasis. They ran a skewness test of the frequency distribution of the observed interactions, and found that synergistic interactions and antagonistic interactions were of approximately the same strength and frequency. They suggested that these results did not support the Deterministic Mutation hypothesis, because if both synergistic and antagonistic interactions are common between deleterious mutations, this would make it more difficult for recombination to evolve (Otto and Feldman, 1977).

De Visser et al. (1997b) obtained results similar to those obtained by Elena and Lenski with *E. coli*. They tested for interactions between a variety of mutations in the filamentous fungus *Aspergillus niger*. Like Elena and Lenski, they also found a log linear relationship between fitness and the number of mutations. When they further analyzed these results, they found that the log linear relationship was the result of a combination of both synergistic and antagonistic epistatic interactions between mutations. These opposing interactions had the effect of partially canceling each other out.

Although these results do not support the Deterministic Mutation hypothesis, it should be pointed out that neither *E. coli* nor *Aspergillus niger* undergo meiotic recombination. It would be useful to have considerably more data concerning epistatic interactions between mutations in sexually recombining organisms before final judgements are made regarding the validity of the Deterministic Mutation hypothesis.

The preceding studies leave the validity of the Deterministic Mutation hypothesis somewhat uncertain. What is needed is more hard data on the frequency of synergistic epistasis in a variety of organisms. In an attempt to determine the best method for collecting such data, West et al. (1998) have recently analyzed three different methods for experimentally testing for synergistic epistasis. They concluded that the most promising method is to cross two different lineages having very different numbers of mutations, and then compare the mean log fitness of the parental lines with the mean log fitness of their offspring. West et al. also reanalyzed the data of De Visser et al. (1996, 1997a), concluding that the "data do not provide clear evidence for or against epistasis between deleterious mutations in *C. moewussii.*

Other models which fall into the Fisher-Muller category distinguished by Felsenstein (1988) include the models of Ghiselin (1974), Williams (1975; Williams and Mitton, 1973), and Strobeck et al. (1976).

A.2 "Vicar Of Bray" or "Fisher-Muller" Type Models in which Environmental Variability Is Viewed as the Driving Force behind the Evolution of Sex

Several of the Vicar of Bray class of theories invoke environmental variability as a possible driving force behind the evolution of sex. Bell (1982) points out that there are two extreme models of environmental variability. The environment can be considered to be coarse-grained and constant in nature, or fine-grained and inconstant. In the first instance, individuals might, during their lifetimes, spend most of their time in a single habitat, but might occasionally stray into a clearly different situation. Although habitats might change in their spatial distribution, the numbers and types of habitat would remain relatively constant over time. Under these circumstances, some form of Tangled Bank situation, involving frequency dependent selection, would be applicable (see below).

In the second instance, in which the environment is fine-grained and inconstant, individuals would experience a wide range of slightly different

TABLE 6. Showing the Relationships Between Various Types of Models, Types of Selection, and Environmental Variation

	Type of Selection	
	Frequency Dependent	Frequency Independent
Environmental Variation in:		
Space	Tangle Bank Sib-Competition	?
Time	Red Queen	Lottery (Best man)

environmental situations during their lifetimes, but overall, at any particular point in time, all individuals would be living under rather similar conditions. At different times during a generation, or from one generation to the next, the entire population may experience different average environments. This type of situation lends itself to a "Best Man" or "Lottery" situation involving frequency-independent selection (Bell, 1982) (see Table 6).

The Tangled Bank Hypotheses Ghiselin (1974) developed a model which attempted to explain the evolution of sex in terms of individual selection, based on spatial, environmental heterogeneity. Bell (1982; 1988b) has given models such as those of Ghiselin the name of "Tangled Bank" hypotheses, from Darwin's metaphor. Such models propose that sex is advantageous in spatially heterogeneous environments in which the environment differs from patch to patch, but in which the type of patch and its frequency remain constant from generation to generation (Bierzychudek (1987); Maynard Smith (1988)). Under these circumstances, sex can be favored because a sexual parent will produce offspring with a variety of different genotypes. Because like genotypes tend to have very similar ecological requirements, and therefore compete more intensely with each other, genetic diversification through sex can be advantageous.

Bell (1985) points out that the agricultural literature is full of examples of mixtures of crops yielding more than the mean of their individual components. Tilman et al. (1996) have recently demonstrated, in a well-replicated field experiment, that both ecosystem productivity and stability increase significantly with plant species diversity. While species diversity is of course a different phenomenon from genetic diversity within a species, these results suggest that at least the premise of the Tangled Bank model has merit.

The advantage is dependent upon a higher diversity in the competing sexual populations than in the asexual ones. It breaks down if the asexual population consists of many genetically distinct clones, each of which is well-adapted to a different niche, for then the sexual population will tend to be outcompeted because it will produce a relatively large number of ill-adapted genotypes each generation (Bell, 1982). Bell examined the plausibility of Tangled Bank models in his book *The Masterpiece of Nature* (1982) and concluded that although the logic was valid, the requirements of the model would rarely be met.

More recently, Burt and Bell (1987) have, after examining the frequency of excess chiasmata in mammals, concluded that the distributions of these frequencies do not support the Tangle Bank model. (For further discussion of their analysis, see the Red Queen section below).

Sib competition models are very similar to Tangled Bank models (Bierzychudek, 1987). In such models, the degree to which siblings compete within a heterogeneous patch depends upon their genetic similarity. Individuals having rare phenotypes are favored because they can take advantage of resources to which few other individuals are adapted, and thus are subjected to less competition than are more common phenotypes. A prediction of this type of model would be that the advantage of sexuality would increase with an increasing number of progeny per parent. Burt and Bell's (1987) observations on excess chiasmata frequency should apply to these models as well as to the Tangled Bank.

"Best Man" or Lottery Models Figure 8 (modified from Bell (1982)) shows the expected differences in fitness of offspring in sexual and asexual populations that each carry the same amount of genetic variation. While the mean fitness of sexual progeny may be somewhat lower than that of asexual progeny, the sexual progeny have a much broader variance in fitness. Even if a majority of the sexual progeny have lower fitnesses than their asexual counterparts, a few of them will be extremely fit. These individuals will contribute disproportionately to the next generation, and this may more than offset any of the costs of sex (Fig. 8).

Lottery models (Williams, 1975; Williams and Mitton, 1973) view sex as being an *individual* reproductive adaptation, allowing a sexual parent to produce offspring with a greater variance in fitness than their asexual counterparts. Some of these offspring will be better able to exploit niches in a temporally changing environment (Bell, 1982). In general, lottery models require extremely strong truncation selection in order for there to be an advantage to sexual recombination (Bell, 1982; Bierzychudek, 1987).

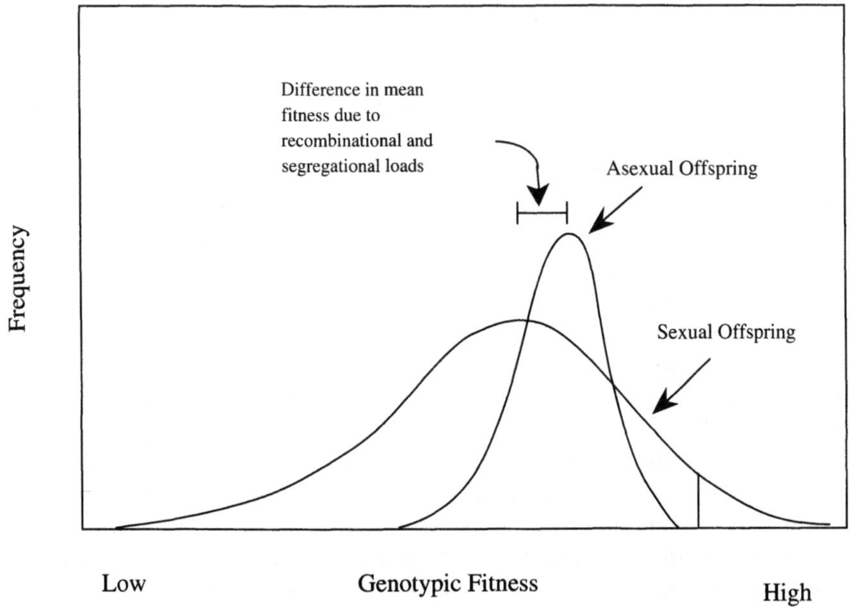

FIG. 8. Showing the variance in fitness of progeny from sexual and asexual parents.

The Hill-Robertson Effect

Hill and Robertson (1966) were the first to describe an important virtually ubiquitous, and yet often unappreciated effect of linkage which has since come to bear their name—the Hill-Robertson effect. Felsenstein (1974; 1988) perhaps best explains this rather abstruse phenomenon by noting the basic principle, that variation from individual to individual in the number of offspring produced, even when this variation is unrelated to any particular genotype, reduces the effective population size (N_e). This in turn reduces the probability of the fixation of an advantageous gene under positive selection. Conversely, it also reduces the effectiveness of selection against deleterious alleles. These two rather opposite outcomes are due to the fact that the likelihood of fixation is dependant upon the produce of $N_e s$, the effective population size and the selection coefficient "s". The greater the variance in reproduction, the smaller the effective size of the population. Felsenstein gives the extreme example of when one individual is responsible for all reproduction, the effective size of the population becomes 1 regardless of the actual population size. Not only can differential reproductive success lead to a reduction in effective population size, but so too can random variances in the fitness of one allele created by linkage to another locus under selection. The tighter the linkage, the more selection at one locus interferes with selection at the other by creating a type of background noise. When there are many loci under selection, and linked to varying degrees, this effect is even more pronounced, making selection at any one locus even less effective. This is the so called Hill-Robertson effect.

Birky and Walsh (1988) have shown that while the Hill-Robertson effect slows the fixation of advantageous mutations, and speeds up the fixation of deleterious mutations, it has no effect on the rate of fixation of neutral mutations. Felsenstein (1988) underlined the importance of this effect when he stated that the Hill-Robertson effect is "the common phenomenon underlying all forms of the Fisher-Muller theory." While we feel that this effect may underlie all of the Vicar of Bray type models, and that it may also contribute significantly to Muller's ratchet, there are other independent processes which can reduce the effective population size and thereby speed up the ratcheting process. These factors would include genetic bottlenecks, unequal mating systems, biased sex ratios, etc.

I. Maintenance of Sex Models which Involve the Removal of Linkage Disequilibrium

B. The Sturtevant-Mather Class of Models

Sturtevant and Mather (1938) postulated a different situation from that envisioned by the constant selection models Fisher and Muller, one in which selection pressures cycle over time. At one point selection will favor coupling gamete configurations (AB and ab) and at another it will favor repulsion configurations (Ab and aB) (Felsenstein, 1988). The easiest way to switch from coupling to repulsion and back again is through some form of sex involving recombination. Thus, recombination should be favored repeatedly as the environment fluctuates. This is in contrast to the Fisher-Muller (Vicar of Bray) model, which favors one type of recombinational event, the one that brings together two advantageous mutations. As Sturtevant and Mather point out, "subsequent recombination would be disadvantageous in breaking up the new favorable combination."

The Red Queen Hypothesis Models falling into the Sturtevant-Mather classification include Red Queen models (Van Valen, 1973), in which interspecific competition and coevolution result in a negative correlation between a genotype's present fitness and its fitness at some time in the past. Red Queen hypotheses often use parasite/host coevolution as examples of such interspecific competition. Under these circumstances, not only will the allele frequencies at loci that determine host resistance (or parasite efficacy) fluctuate through time, but so too will the coefficient of linkage disequilibrium, cycling from positive to negative and back (Hamilton et al., 1990). The importance of sex in such a model is that "it stores genes that are currently bad but have promise for reuse at some time in the future. It continually tries them in combination, waiting for the time when the focus of disadvantage has moved elsewhere. When this has happened, the genotypes carrying such genes spread by successful reproduction, becoming simultaneously stores for other 'bad' genes and thus onward in a continuous succession." (Hamilton et al., 1990).

Asexual organisms have no such mechanism by which to store currently "bad" alleles for future use. Once these alleles are lost new similar alleles may arise very slowly by mutation, or they may not arise at all. Hamilton et al. make use of a simple analogy to help in visualizing their model. Suppose a ball floating on water has a cube inside it, each vertex of which is just touching the surface of the ball. Now suppose that at each of the eight vertices of the cube a single genotype colony of the host species grows. Each vertex supports a unique genotype which, for a two-allele, three-locus haploid host, can be labeled in the following manner: 000, 001, 010, 100, 011, 101, 110, and 111. As these colonies grow, the weight of the largest colony causes the ball to roll toward the vertex at which this colony is growing. Eventually this colony will become submerged and will drown. Now the ball rolls in the direction of the next largest colony and the process is repeated. The weight of the colonies in this model does not have to do with mass, but rather with "an accumulation of ill-health that follows upon abundance," due to coevolution by the parasite.

It can be seen that once a genotype is lost, it can be recreated in sexual hosts by recombination between genotypes at other vertices. This is not the case for an asexual host. Once a genotype becomes submerged, there is no way to recreate it except perhaps by mutation, which would be a very slow process. Hamilton et al. (1990) point out that as the number of polymorphic loci determining host resistance increases, the likelihood that any particular allele will be lost decreases in the sexual population. This is because each new two-allele locus added to the host defense system doubles the previous number of genotypes (vertices) at which the carriers of an allele can escape destruction.

There is some evidence supporting the Red Queen hypothesis for the maintenance of sex. Burt and Bell (1987) undertook an extensive study of excess chiasma frequencies (defined as "the total number of chiasmata in excess of one per bivalent") in mammalian sperm in an attempt to distinguish between the hypotheses of the Tangled Bank (involving *spatial* heterogeneity of the *physical* environment), and the Red Queen (involving *temporal* heterogeneity of the *biotic* environment). According to the Red Queen hypothesis, the host species with longer generation times are expected to be engaged in a more vigorous adaptive race with their parasites. This is because their parasites have more time in each generation to evolve counteradaptations to the hosts' defenses. One prediction therefore is that hosts with longer generation times would recombine more during meiosis than would those with shorter generations.

The Tangled Bank, on the other hand, would predict no correlation between generation time and recombination, but rather a positive correlation between the average number of offspring produced and the amount of recombination (Burt and Bell, 1987). This is because the greater the number of progeny produced, the greater the potential for competition between siblings that resemble each other. If the progeny are more diverse because of high levels of recombination, such competition will be reduced.

Mammals were chosen in this survey because age to maturity is negatively correlated with litter size, and therefore one of the two hypotheses can be rejected. Burt and Bell found that the correlation between excess chiasmata and generation time was very strong ($r = 0.875$, $p < 0.001$, $N = 24$), while the correlation between excess chiasmata and litter size was actually negative and not as significant ($r = -0.504$, $p < 0.01$, $N < 424$). Burt and Bell concluded that "crossing-over may function to combat antagonists with short generation times but does not function to reduce sib competition."

In another paper, Lively et al. (1990) reported the results of a study of parasite loads in sexual and asexual populations of *Poeciliopsis* co-inhabiting ponds in Sonora, Mexico. Their results also appear to support the Red Queen hypothesis. In a pond in which there was a genetically diverse sexual population and an asexual clonal population, the clonal population had significantly more parasites per individual than the sexual population. In another pond in which there were two clonal populations, one more common than the other, in addition to a sexual population, the common clonal population was found to have a significantly higher parasite load than either of the other populations. Finally, in a third pool in which a genetically depauperate sexual population (resulting from an extinction and subsequent founding event) and a clonal population coexisted, the sexual population was found to have a significantly higher parasite load than the asexual population. A manipulation experiment was conducted in which sexual females from another pond were introduced to this pond, thereby restoring genetic diversity. Two years later it was found that the now genetically diverse sexual population had a significantly lower parasite load than the asexual clone.

Moritz et al. (1991), studying mite infestations of parthenogenetic and sexual forms of the lizard *Heteronotia binoei*, found parthenogenetic forms to be far more susceptible to the effects of parasitism than their sexual counterparts. This relationship held true in several different localities that had genetically different clonal and sexual forms, in spite of the fact that the parthenogenetic forms were found to be genetically diverse.

Clarke et al. (1994) found evidence supporting at least one aspect of the Red Queen. They ran long-term, paired competitions between strains of vesicular stomatitis virus (VSV) of approximately equal initial relative fitness. Because VSV is an RNA virus having a very high mutation rate, it is able to evolve rapidly even in the absence of sex. Competing strains were transferred repeatedly, being allowed to grow and compete between transfers. The relative proportion of each competing strain was monitored at intervals. It was found that over the first 12–15 passages competing strains maintained very similar relative fitnesses. Then, rather suddenly, one strain would rapidly increase, displacing the other. The relative fitnesses of the winning strains were found to have increased significantly by the end of the competitions. More surprisingly, it was found that the fitness of the losing strains had also increased! This meant that both the competing strains had undergone a Red Queen coevolution.

More recent data in support of the Red Queen has come from a five year study of a New Zealand fresh water snail (*Potamopyrgus antipodarum*) and its trematode parasites (*Microphallus spp.*) (Dybdahl and Lively, 1998). *P. antipodarum* exists in two different forms: a sexual diploid form, or an asexual triploid form (Dybdahl and Lively, 1995). All of the snails in this study were apomictic asexuals from a single lake on the south island, in which sexual strains were absent. The study of these asexual clones was aided by the fact that allozyme markers identifying specific clones are linked to the loci determining resistance to trematode infection.

Although more than a hundred clones were identified over the five year study period, no more than four clones were common in any one year. The remainder of the lake's snail population consisted of an array of rare clones. Over the five year study, Dybdahl and Lively tracked the frequency of the four most common clones along with their level of infection. What they found was that the host clonal frequencies cycled over time, accompanied by a time-lagged cycling of parasite super-infection. This super-infection subsequently led to a decrease in the frequency of the common clone. In other words, as a clone increased in frequency, the level at which individuals of this clone were infected also increased, although in a somewhat delayed manner. After the level of infection had increased the frequency of that particular clone was driven down (presumably due to this super-infection).

In order to test whether recently common clones were more easily infected than rarer clones, Dybdahl and Lively conducted a laboratory experiment, exposing different clones of snail to *Microphallus* infection. They found that rarer clones showed significantly less infection than common clones ($X^2 = 41.391$, df = 1, $p < 0.0001$). In a subsequent set of lab experiments, (unpublished data), they showed that while rare and common clones showed differential susceptibilities to infection from parasites cohabiting the same lake, both types of clone were equally susceptible to infection from parasites inhabiting a different lake.

While Dybdahl and Lively's (1998) study involved only asexual clones, the implications of their results for the Red Queen hypothesis and the maintenance of sexual recombination are clear. Not only did they demonstrate that parasites could drive host populations through cycles, effectively "tracking" the most common host genotypes, but they also demonstrated an advantage for the rare genotype. Such rare genotypes are constantly being created in sexual organisms through sexual recombination.

Not all studies involving parasite loads in sexual and asexual species support the Red Queen hypothesis. Hanley et al. (1995) found, in a study of mite infestations of a parthenogenetic gecko species and its two sexual ancestors, that the asexual species had significantly lower infestation rates than either of its syntopic sexual ancestors. These authors suggest that heterosis resulting from the recent hybrid origin of the asexual species may be at least partially responsible, but point out that in the similar study by Moritz et al. (1991) of sexual and asexual *Heteronotia* (the latter also being of a hybrid origin), it was the sexuals that showed a lower level of parasite infestation. Resolution of such conflicts will require careful measurements of the genetic variation that is actually involved in parasite resistance and susceptibility, which of course may be different from the levels of isozyme or DNA variability that are normally measured in population surveys.

II. Maintenance of Sex Models which Do Not Involve the Removal of Linkage Disequilibrium

Theories which do not attempt to explain sex as an adaptation to reduce linkage disequilibrium, and which therefore fall into Felsenstein's second category, include the *DNA repair hypotheses* (Bernstein, 1979; Bernstein, 1987; Bernstein and Bernstein, 1991; Bernstein and Chen, 1987; Bernstein and Johns, 1989; Bernstein et al., 1984; Bernstein et al., 1985b; Bernstein et al., 1985c; Bernstein et al., 1987; Bernstein et al., 1988; Bernstein et al., 1981; Walker 1978; Bengtsson, 1985), the *Contagion hypothesis* (Hickey and Rose, 1988; Hickey, 1982), and the *DNA methylation hypothesis* (Holliday, 1984; Holliday, 1988; Holliday, 1989a; Holliday, 1989b). In this review, the first two categories of hypotheses have been grouped perhaps more appropriately under hypotheses concerned with the origin of sex and were dealt with in that section. This leaves only the methylation hypothesis to be dealt with here.

The DNA Methylation Hypothesis

Robin Holliday's DNA methylation model (Holliday, 1984; 1988; 1989a; 1989b) attempts among other things to explain the maintenance of sex in higher eukaryotic organisms. It is similar to Bernstein's model in that it views meiosis as a means of repairing damage. The primary difference between these two models is that in Holliday's model the damage is epigenetic rather than genetic in the usual sense. Epigenetic damage is damage that affects the "unfolding of the developmental program" during ontogeny (Holliday, 1989a). According to the Holliday model, the only means of repairing this epigenitic damage is through meiosis.

A great deal of evidence has been accumulated over the past two decades showing the importance of DNA methylation in the regulation of gene expression (Cattanach and Kirk, 1985a; Cedar, 1988; Doerfler et al., 1989; Hadchouel et al., 1987; Hare and Taylor, 1989; Holliday, 1984; Holliday, 1987; Holliday, 1988; Holliday, 1989a; Holliday, 1989b; Jahner et al., 1982; Kruger et al., 1989; McGrath and Solter, 1984; Monk, 1986; Monk, 1987; Niwa and Sugahara, 1981; Nur et al., 1989; Reik et al., 1987; Riggs, 1989; Surani et al., 1988; Surani et al.,1984; Surani et al., 1987; Swain et al., 1987; Wilson and Jones, 1983). In higher eukaryotes, genes that are highly methylated are generally repressed at the transcriptional level, while if the same genes lose their methylation they are derepressed and become transcriptionally active. A number of lines of evidence suggest and important role for DNA methylation in many instances of the phenomenon known as genomic imprinting (Li et al., 1993; Zuccotti and Monk, 1995; Tremblay et al., 1995).

Genomic Imprinting

Genomic imprinting is defined as the differential expression of maternally derived and paternally derived genes (Crouse, 1960). Generally, it is the imprinted allele that is transcriptionally inactive (Bartolomei and Tilghman, 1997). While the term 'imprinting' implies differential expression of maternally derived and paternally derived genes, it embraces other mechanisms in addition to methylation. Imprinting has been found in plants (for a review see Matzke and Matzke, 1993) animals (for a review see Bartolomei and Tilghman, 1997), and fungi (Klar and Bonaduce, 1993). It has been shown, at least in some mammals, that paternally and maternally derived gametes are differentially imprinted during gametogenesis, and that both complementary types of imprinting are required for normal development of the embryo (McGrath and Solter, 1984; Surani et al., 1984; Barton et al., 1984). This imprinting appears to be reprogrammed during meiosis. Once a zygote is formed however, the original paternal and maternal imprints may become highly modified in a pre-programmed fashion during the course of development. Although the somatic cells of an adult organism carry both a modified paternally imprinted haploid complement and a modified maternally imprinted haploid complement, the effects of this previous generation of imprinting are cancelled during meiosis. All of the meiotic products from such an individual will have either a fresh "paternal imprint" or a fresh "maternal imprint," depending upon the sex of the individual (Holliday, 1989a).

Holliday has suggested that "epigenetic defects" such as the loss of methylation from a regulatory region of a gene could be responsible for a variety of problems including aging of the soma and cancer. According to his model, meiosis is crucial to the preservation of the integrity of the germ line because it allows the correction of such epigenetic defects. However, while meiosis may indeed be required in order to carry out fresh genomic imprinting and in order to repair epigenetic defects, it seems highly implausible that meiosis is maintained solely for its role in imprinting in certain organisms.

CONCLUDING REMARKS

The many theories concerning the origin and maintenance of sexual reproduction have often been presented in an either-or fashion. We consider, however, that there is no one correct answer to the question "Why does sex exist?" Rather, for each of the numerous types of sex-like phenomena, there are a number of probable explanations, and several different processes may be operating simultaneously. We wish to espouse the view that all of the models discussed in this review have a degree of validity. No one model is correct to the exclusion of the other models. For example, while there is a great deal of evidence in support of the contagion hypothesis for the origin of sex, this need not imply that DNA repair, or even conspecific cannibalism, were not also involved in its origin. It seems probable that recombination mechanisms did evolve in order to cope with DNA damage. There is no reason to believe that these same repair mechanisms could not at some later date be co-opted by selfish genetic parasites and used by these elements to propagate themselves. These genetic parasites themselves could at some latter date be co-opted by the "host" organism for its own use and benefit, as appears to be the case with the mating type idiomorphs of the bipolar fungi.

While the contagion hypothesis may in part explain the origin of sex, it seems unlikely that it can explain its maintenance. Once sex had evolved, it seems certain that other forces would take over its maintenance.

As with the origin of sex, the maintenance of sex can be attributable to a number of forces, some of which play a more important role in some organisms than others, and in some environments more than others. For instance, while sexual recombination may indeed serve to reduce sibling competition by reducing the genetic correlation between them, it seems likely that this is of second or third order importance in comparison to the benefits afforded by sexual recombination with regards to the repair of DNA damage, and the removal of deleterious mutations, and perhaps the evasion of one's parasites. Similarly, while concatenation of beneficial mutations surely is a fortuitous byproduct of sex, it too must take a back seat to the overwhelming effects of parasites, deleterious mutations, and DNA damage. There is absolutely no reason to believe that Muller's ratchet and the deterministic mutation hypothesis don't both operate to varying degrees in different organisms. Clearly, in organisms such as *Saccharomyces* and *C. elegans* which have low genomic mutation rates, the deterministic mutation hypothesis can not explain the maintenance of sex, while Muller's ratchet may in part. On the other hand, in organisms with high genomic mutation rates, the determinist mutation hypothesis may very well be the predominate reason for the maintenance of sex.

DNA damage is pervasive, occurring in all organisms, and hence, so too is the need to be able to cope with such damage. While DNA repair has been invoked to explain the origin of sexual recombination, DNA repair may also be a leading force responsible for the maintenance of sex once it arose. It is worth reiterating the distinction between DNA *damage*, and *mutation*, both of which exert powerful selective pressures for the maintenance of sexual recombination. The former is a physical alteration in the normal conformation of the DNA, which if not repaired can lead to death of the organism. Being a physical alteration, damage can be recognized by repair enzymes and dealt with. Mutations, on the other hand, cannot be recognized by repair enzymes. They are simply a replacement of one nucleotide by another, and as such, the only means by which they can be "recognized" is by exposing them to selection. Thus, while sexual recombination can combat both DNA damage as well as DNA mutation, it does so by different means. In the case of damage, meiotic recombination allows the use of a homologous piece of DNA as a template for the recombinational repair of the damaged section. In the case of mutation, sexual recombination provides two distinct means by which to combat the effects of deleterious mutations. The first of these is the ability to recreate mutationally less loaded lineages from lineages having greater mutational loads (i.e. combating Muller's ratchet). The second is the ability to increase the variance in the number of mutations per individual following truncation like selection. This allows a number of mildly deleterious mutations to be brought into a single individual which truncation selection can then eliminate from the population (Kondrashov's Deterministic Mutation Model).

While mutation and DNA damage may impose the most immediate selective pressures for the maintenance of sex, temporal changes in the biotic environment may not be too far behind. It seems likely that the Red Queen model plays an important role in the struggle of many organisms to keep pace with their biotic environment. Dybdahl and Lively (1998) have recently provided important data in support of this notion. What is needed now is more data from a variety of different organisms in order to help illuminate the ubiquity of processes such as the Red Queen.

Some of the hypotheses no doubt play very limited (although perhaps critical) role in the maintenance of sex. The DNA methylation hypothesis, for instance, can only apply to certain classes of sexual organism (i.e. those possessing differential methylation patterns in paternally derived and maternally derived gametes). It is has been shown in mammals with genetic imprinting, that in order for a zygote to develop normally, both complementary paternally imprinted and maternally imprinted gametic nuclei are required. If both imprints are not present, the zygote will not survive. This effectively makes bi-parental reproduction obligatory.

While genomic imprinting may play a role in genomic defense against selfish genetic elements, the fact that not all organisms possess imprinting demonstrates once again the multifaceted nature of the forces responsible for the maintenance of sexual recombination, as well as the limited degree to which any one force may be able to explain sex in all sexual organisms.

GLOSSARY OF TERMS

Name of Process	**Description**
Amphimixis (Mixis)	Syngamy or karyogamy leading to fertilization to form an individual with two different parents; outcrossing, outbreeding (Margulis and Sagan, 1984). The life cycle consists of an alternation of syngamy and meiosis (Kondrashov, 1993).

Apomixis	Altered meiosis or fertilization such that amphimixis is bypassed (e.g., parthenogenesis).
Automixis	Syngamy or karyogamy of nuclei or cells derived from the same parent (selfing; extreme inbreeding; autogamy).
Autopoiesis	Metabolic self-maintenance at the expense of carbon and energy sources. A characteristic of all living cells, but not of viruses.
Crossingover	Breakage and reunion of DNA of homologous non-sister chromatids during meiosis or mitosis.
Coupling Gamete (see also Repulsion Gamete)	A gamete having alleles at two or more loci which are designated to be alike in some manner. (e.g. AB and ab are coupling configurations) (Futuyma, 1986).
Epigenetic	Developmental; pertaining especially to interactions among developmental processes above the level of primary gene action (Futuyma, 1986).
Epistasis	Any departure from multiplicative fitness interactions between mutations on a normal scale, or additive interactions between mutations on a log scale).
Idiomorphs	Alleles occupying homologous positions on the chromosome but having such dissimilar sequences that they do not have a recent common ancestor (Metzenberg and Glass, 1990).
Karyogamy	Fusion of gamete nuclei.
Karyotype	The chromosome complement of an individual.
Linkage	Tendency for two loci on the same chromosome to segregate together.
Linkage Equilibrium/ Disequilibrium	If alleles at two or more loci are associated more or less frequently than predicted by their individual frequencies, they are in linkage disequilibrium; if not, they are in linkage equilibrium (Futuyma, 1986).

Locus	The specific site on a chromosome occupied by a gene; more loosely, the gene itself, in all its allelic states (Futuyma, 1986).
Mixis (sexuality)	"The rearrangement of genetic material through meiosis or syngamy or (usually) both, almost always resulting in the production of one or more new organisms differing genetically from one another and from their parents." (Bell, 1982).
Parthenogenesis	Development of eggs or macrogametes in the absence of amphimixis.
Pleiotropy	The phenotypic effect of a gene on more than one characteristic (Futuyma, 1986).
Recombination	Breakage and reunion of DNA molecules (Margulis, Sagan, and Olendzenski, 1985). More broadly, any process resulting in hybrid DNA molecules.
Reproduction	The process that augments the number of cells or organisms.
Repulsion Gamete	A gamete bearing at two or more loci alleles that are different in some way. e.g. Ab and aB are repulsion configurations (Futuyma, 1986).
	Any process that unites genes (DNA) in an individual cell or organism from more than a single source (Margulis, et al., 1985).
Sex	"A process in which (all or part of) the genomes of two parents are brought together in a common cytoplasm to produce progeny which may then contain reassorted portions of the parental genomes" (Bernstein, Byerly, Hopf, and Michod, 1984b).
Syngamy	Contact or fusion of gametes (cells).

REFERENCES

Achtman, M., Azuma, T., Berg, D. E., Ito, Y., Morelli, G., Pan, Z.-J., Suerbaum, S., Thompson, S. A., van der Ende, A., and van Doorn, L.-J., 1999, Recombination and clonal groupings within *Helicobacter pylori* from different geographical regions, *Mol. Microbiol.* **32:**459–470.

Adams, M. D., Celniker, S. E., Holt, R. A., Evans, C. A., Gocayne, J. D., Amanatides, P. G., Scherer S. E., Li, P. W., Hoskins, R. A., Galle, R. F., George, R. A., Lewis, S. E., Richards, S., Ashburner, M., Henderson, S. N., Sutton, G. G., Wortman, J. R., Yandell, M. D., Zhang, Q., Chen, L. X., Brandon, R. C., Rogers, Y. H., Blazej, R. G., Champe, M., Pfeiffer, B. D., Wan, K. H., Doyle, C., Baxter, E. G., Helt, G., Nelson, C. R., Gabor Miklos, G. L., Abril, J. F., Agbayani, A., An, H. J., Andrews-Pfannkoch, C., Baldwin, D., Ballew, R. M., Basu, A., Baxendale, J., Bayraktaroglu, L., Beasley, E. M., Beeson, K. Y., Benos, P. V., Berman, B. P., Bhandari, D., Bolshakov, S., Borkova, D., Botchan, M. R., Bouck, J., Brokstein, P., Brottier, P., Burtis, K. C., Busam, D. A., Butler, H., Cadieu, E., Center, A., Chandra, I., Cherry, J. M., Cawley, S., Dahlke, C., Davenport, L. B., Davies, P., de Pablos, B., Delcher, A., Deng, Z., Mays, A. D., Dew, I., Dietz, S. M., Dodson, K., Doup, L. E., Downes, M., Dugan-Rocha, S., Dunkov, B. C., Dunn, P., Durbin, K. J., Evangelista, C. C., Ferraz, C., Ferrier, S., Fleischmann, W., Fosler, C., Gabrielian, A. E., Garg, N. S., Gelbart, W. M., Glasser, K., Glodek, A., Gon, G. F., Gorrell, J. H., Gu, Z., Guan, P., Harris, M., Harris, N. L., Harvey, D., Heiman, T. J., Hernandez, J. R., Houck, J., Hostin, D., Houston, K. A., Howland, T. J., Wei, M. H., Ibegwam, C., Jalali, M., Kalush, F., Karpen, G. H., Ke, Z., Kennison, J. A., Ketchum, K. A., Kimmel, B. E., Kodira, C. D., Kraft, C., Kravitz, S., Kulp, D., Lai, Z., Lasko, P., Lei, Y., Levitsky, A. A., Li, J., Li, Z., Liang, Y., Lin, X., Liu, X., Mattei, B., McIntosh, T. C., McLeod, M. P., McPherson, D., Merkulov, G., Milshina, N. V., Mobarry, C., Morris, J., Moshrefi, A., Mount, S. M., Moy, M., Murphy, B., Murphy, L., Muzny, D. M., Nelson, D. L., Nelson, D. R., Nelson, K. A., Nixon, K., Nusskern, D. R., Pacleb, J. M., Palazzolo, M., Pittman, G. S., Pan, S., Pollard, J., Puri, V., Reese, M. G., Reinert, K., Remington, K., Saunders, R. D., Scheeler, F., Shen, H., Shue, B. C., Siden-Kiamos, I., Simpson, M., Skupski, M. P., Smith, T., Spier, E., Spradling, A. C., Stapleton, M., Strong, R., Sun, E., Svirskas, R., Tector, C., Turner, R., Venter, E., Wang, A. H., Wang, X., Wang, Z. Y., Wassarman, D. A., Weinstock, G. M., Weissenbach, J., Williams, S. M., Woodage, T., Worley, K. C., Wu, D., Yang, S., Yao, Q. A., Ye, J., Yeh, R. F., Zaveri, J. S., Zhan, M., Zhang, G., Zhao, Q., Zheng, L., Zheng, X. H., Zhong, F. N., Zhong, W., Zhou, X., Zhu, S., Zhu, X., Smith, H. O., Gibbs, R. A., Myers, E. W., Rubin, G. M., and Venter, J. C., 2000, The genome sequence of *Drosophila melanogaster*, *Science* **287:**2185–2195.

Andersson, D. I., and Hughes, D., 1996, Mullér's ratchet decreases fitness of a DNA-based microbe, *Proc. Natl. Acad. Sci. USA*, **93:**906–907.

Akrigg, A., and Ayad, S. R., 1970, Studies on the competence-inducing factor of *Bacillus subtilis*, *Biochem. J.*, **117:**397–403.

Akrigg, A., Ayad, S. R., and Barker, G. R., 1967, The nature of a competence-inducing factor in *Bacillus subtilis*, *Biochemical And Biophysical Research Communications*, **28:** 1062–1067.

Albritton, W. L., Setlow, J. K., Thomas, M., Sottnek, F., and Steigerwalt, A. G., 1984, Heterospecific transformation in the genus *Haemophilus*, *Molecular And General Genetics*, **193:**358–363.

Andersson, D. I., and Hughes, D., 1996, Muller's ratchet decreases fitness of a DNA-based microbe. *Proc. Natl. Acad. Sci. USA*, **93:**906–907.

Anker, P., Stroun, M., Gahan, P., Rossier, A., and Greppin, H., 1971, Natural release of bacterial nucleic acids into plant cells and crown gall induction, in: (L. Ledoux, ed.), *International Symposium On Uptake Of Informative Molecules By Living Cells*, (pp. 193–200). American Elsevier Publishing, Inc., New York.

Bartolomei, M. S., and Tilghman, S. M., 1997, Genomic imprinting in mammals, *Ann. Rev. Genet.*, **31:**493–525.

Barton, S. C., Surani, M. A. H., and Norris, M. L., 1984, Role of paternal and maternal genomes in mouse development, *Nature*, **311:**374–376.
Bataillon, T., 2000, Estimation of spontaneous genome-wide mutation rate parameters: whither beneficial mutations? *Heredity*, **84:**497–501.
Bateman, A. J., 1959, The viability of near-normal irradiated chromosomes, *Internat. J. Rad. Biol.* **1:**170–180.
Beach, D. H., and Klar, A. J., 1984, Rearrangements of the transposable mating-type cassettes of fission yeast, *EMBO J.*, **3:**603–610.
Behnke, D., 1981, Plasmid transformation of *Streptococcus sanguis* (Challis) occurs by circular and linear molecules, *Mol. Gen. Genet.*, **181:**490–497.
Bell, G., 1982, *The Masterpiece of Nature: the evolution and genetics of sexuality*, University of California Press, Los Angeles.
Bell, G., 1985, Two theories of sex and variation, *Experientia*, **41:**1235–1245.
Bell, G., 1988a, *Sex and Death in Protozoa: The History of an Obsession*, Cambridge University Press, Cambridge.
Bell, G., 1988b, Uniformity and diversity in the evolution of sex, in *The Evolution of Sex*, (R. E. Michod and B. R. Levin, eds.), pp. 126–138, Sinauer Associates, Inc., Sunderland, Mass.
Bell, G., 1993, The sexual nature of the eukaryote genome, *J. Hered.*, **84:**351–359.
Bengtsson, B. O., 1985, Biased gene conversion as the primary function of recombination, *Genetical Research*, **47:**771–780.
Bengtsson, B. O., 1990, The effect of biased conversion on the mutation load, *Genet. Res.*, **55:**183–187.
Bengtsson, B. O., 1992, Deleterious mutations and the origin of the meiotic ploidy cycle, *Genetics*, **131:**741–744.
Bernstein, C., 1979, Why are babies young? *Perspect. Biol. Med.*, **22:**539–544.
Bernstein, C., 1987, Damage in DNA of an infecting phage T4 shifts reproduction from asexual to sexual allowing rescue of its genes, *Genet. Res.*, **49:**183–189.
Bernstein, C., and Bernstein, H., 1991, *Aging, Sex, and DNA repair*, Academic Press, San Diego
Bernstein, C., and Chen, D., 1987, Recombinational repair of hydrogen-peroxide induced damages in DNA of phage T4, *Mut. Res.*, **184:**87–98.
Bernstein, C., and Johns, V., 1989, Sexual reproduction as a response to H_2O_2 damage in *Schizosaccharomyces pombe*, J. Bact., **171:**1893–1897.
Bernstein, H., Byerly, H., Hopf, F., and Michod, R., 1985a, DNA repair and complementation: The major factors in the origin and maintenance of sex, in: *The Origin and Evolution of Sex* (H. O. Halvorson and A. Monroy, eds.) Vol. 7 pp. 29–45, Marine Biological Laboratory, Woods Hole, Massachusetts: Alan R. Liss Inc., New York.
Bernstein, H., Byerly, H. C., Hopf, F. A., and Michod, R. E., 1984, Origin of sex. *J. Theor. Bio.*, **110:**323–351.
Bernstein, H., Byerly, H. C., Hopf, F. A., and Michod, R. E., 1985b, The evolutionary role of recombinational repair and sex, *Int. Rev. Cyt.*, **96:**1–28.
Bernstein, H., Byerly, H. C., Hopf, F. A., and Michod, R. E., 1985c, Gentic damage, mutation, and the evolution of sex, *Science*, **229:**1277–1281.
Bernstein, H., Byerly, H. C., Hopf, F. A., and Michod, R. E., 1985d, Sex and the emergence of species, *J. Theor. Bio.*, **117:**665–690.
Bernstein, H., Byerly, H. C., Hopf, F. A., and Michod, R. E., 1987, The molecular basis of the evolution of sex, *Adv. Genet.*, **24:**323–370.
Bernstein, H., Byerly, H. C., Hopf, F. A., and Michod, R. E., 1988, Is meiotic recombination an adaptation for repairing DNA, producing genetic variation, or both? in: *The Evolution of Sex* (R. E. Michod and B. R. Levin, eds.), pp. 139–160, Sinauer Associates, Sunderland.

Bernstein, H., Byers, G. S., and Michod, R. E., 1981, Evolution of sexual reproduction: importance of DNA repair, complementation, and variation, *Am. Nat.*, **117:**537–549.

Bertani, G., and Baresi, L., 1987, Genetic transformation in the methanogen *Methanococcus voltae* PS, *J. Bact.*, **169:**2730–2738.

Bierzychudek, P., 1987, Resolving the paradox of sexual reproduction: A review of experimental tests, in: *The Evolution of Sex and its Consequences* (S. C. Stearns, eds.), pp. 163–174, Sinauer Associates, Sunderland.

Birdsell, J., and Wills, C., 1996, Significant competitive advantage conferred by meiosis and syngamy in the yeast Saccharomyces cerevisiae, *Proc. Natl. Acad. Sci. USA*, **93:**908–912.

Birky, C. W. Jr., 1978, Transmission genetics of mitochondria and chloroplasts, *Ann. Rev. Genet.* **12:**471–512.

Birky, C. W., and Walsh, J. B., 1988, Effects of linkage on rates of molecular evolution, *Proc. Natl. Acad. Sci., USA*, **85:**6414–6418.

Bodmer, W. F., 1972, The evolution of recombination mechanisms in bacteria, in: *Uptake Of Informative Molecules By Living Cells* (L. Ledoux, ed.), pp. 130–139, North-Holland Publishing Company, American Elsevier Publishing Company, Amsterdam, London, New York.

Bowring, F. J., and Catchesise, D. E. A., 1996, Gene conversion alone accounts for more than 90% of recombination events at the *am* locus of *Neurospora crassa*, *Genetics*, **143:**129–136.

Bresler, S. E., 1975, Theory of misrepair mutagenesis, *Mut. Res.*, **29:**467–472.

Burt, A., and Bell, G., 1987, Mammalian chiasma frequencies as a test of two theories of recombination, *Nature*, **326:**803–805.

Butcher, D., 1995, Muller's ratchet, epistasis and mutation effects, *Genetics*, **141:**431–437.

Carlson, C. A., Pierson, L. S., Rosen, J. J., and Ingraham, J. L., 1983, *Pseudomonas stutzeri* and related species undergo natural transformation, *J. Bact.*, **153:**93–99.

Catlin, W., 1956, Extracellular deoxyribonucleic acid of bacteria and a deoxyribonuclease inhibitor, *Science* **124:**441–442.

Catlin, B. W., 1960, Transformation of *Neisseria meningitidis* by deoxyribonucleates from cells and from culture slime, *J. Bact.*, **79:**579–590.

Cattanach, B. M., and Kirk, M., 1985a, Differential activity of maternally and paternally derived chromosome regions in mice, *Nature*, **315:**496–498.

Cedar, H., 1988, DNA methylation and gene activity. *Cell*, **53:**3–4.

Chao, L., 1990, Fitness of RNA viruses decreased by Muller's ratchet, *Nature*, **348:**454–455.

Chao, L., Tran, T., and Matthews, C., 1992, Muller's ratchet and the advantage of sex in the RNA virus ϕ6, *Evolution*, **46:**289–299.

Charlesworth, B., 1978, Model for evolution of Y chromosomes and dosage compensation, *Proc. Natl. Acad. Sci. USA*, **75:**5618–5622.

Charlesworth, D., and Charlesworth, B., 1995, Quantitative genetics in plants: the effect of the breeding system on genetic variability, *Evolution* **49:**911–920.

Charlesworth, D., Morgan, M. T., and Charlesworth, B., 1993, Mutation accumulation in finite outbreeding and inbreeding populations, *Genet. Res.*, **61:**39–56.

Chauvat, F., Astier, C., Vedel, F., and Joset-Espardellier, F., 1983, Transformation in the cyanobacterium *Synechococcus* R2: improvement of efficiency; role of the pUH24 plasmid, *Mol. Gen. Genet.*, **191:**39–45.

Chovnick, A., Ballantyne, G. H., Baillie, D. L., and Holm, D. G., 1970, Gene conversion in higher organisms: Half-tetrad analysis of recombination within the rosy cistron of *Drosophila melanogaster*, *Genetics*, **66:**315–329.

Clark, A. G., and Wang, L., 1997, Epistasis in measured genotypes: *Drosophila* P-element insertions, *Genetics*, **147:**157–163.

Clarke, D. K., Duarte, E. A., Elena, S. F., Moya, A., Domingo, E., and Holland, J., 1994, The red queen reigns in the kingdom of RNA viruses, *Proc. Natl. Acad. Sci. USA*, **91:**4821–4824.

Cleveland, L. R., 1947, The origin and evolution of meiosis, *Science*, **105:**287–288.

Courtois, J., Courtois, B., and Guillaume, J., 1988, High-frequency transformation of *Rhizobium meliloti*, *J. Bact.*, **170:**5925–5927.

Cox, E. C., and Gibson, T. C., 1974, Selection for high mutation rates in chemostats, *Genetics*, **77:**169–184.

Crouse, H., 1960, The controling element in sex chromosome behavior in *Sciara*, *Genetics*, **45:**1429–1443.

Crow, J., and Kimura, M., 1965, Evolution in sexual and asexual populations, *Am. Nat.*, **XCIX:**439–450.

Crow, J., and Kimura, M., 1969, Evolution in sexual and asexual populations: a reply, *Am. Nat.* **103:**89–91.

Crow, J. F., and Kimura, M., 1970, *An Introduction To Population Genetics Theory*, Harper and Row, New York.

Cummings, D. J., MacNeil, I. A., Domenico, J., and Matsuura, E. T., 1985, Excision-amplification of mitochondrial DNA during senescence in *Podospora anserina.* DNA sequence analysis of three unique "plasmids", J. Mol. Bio. **185:**659–680.

Cupples, D. A., Van Etten, J. L., Burbank, D. E., Lane, L. C., and Vidaver, A. K., 1980, In vitro translation of three bacteriophage ϕ6 RNAs, *J. Virology*, **35:**249–251.

Daly, M. J., and Minton, K. W., 1995, Interchromosomal recombination in the extremely radioresistant bacterium *Deinococcus radiodurans*, *J. Bact.*, **177:**5495–5505.

Davies, E. K., Peters, A. D., and Keightley, P. D., 1999, High frequency of cryptic deleterious mutations in *Caenorhabditis elegans.*, *Science* **285:**1748–1751.

Davies, P. J., Evans, W. E., and Parry, J. M., 1975, Mitotic recombination induced by chemical and physical agents in the yeast *Saccharomyces cerevisiae*, *Mut. Res.*, **29:**301–314.

Dee, J., 1982, Genetics of *Physarum polycephalum*, in: *Cell Biology Of Physarum and Didymium* (J. W. Daniel, ed.), pp. 211–251, Academic Press, New York.

Deng, H.-W., and Lynch, M., 1996, Estimation of deleterious-mutation parameters in natural populations, *Genetics* **144:**349–360.

Deng, H.-W., and Lynch, M., 1997, Inbreeding depression and inferred deleterious mutation parameters in *Daphnia*, *Genetics* **147:**147–155.

Dennis, E. S., and Brettell, R. I. S., 1990, DNA methylation of maize transposable elements is correlated with activity, *Phil. Trans. R. Soc. Lond.*, *B*, **326:**217–229.

Denver, D. R., Morris, K., Lynch, M., Vassilieva, L., and Thomas, W. K., 2000, High direct estimate of the mutation rate in the mitochondrial genome of *Caenorhabditis elegans*, *Science* **289:**2342–2344.

De Visser, J. A. G. M., Hoekstra, R. F., and van den Ende, H., 1996, The effect of sex and deleterious mutations on fitness in *Chlamydomonas Proc. R. Soc. Lond.*, *B*, **263:**193–200.

De Visser, J. A. G. M., Hoekstra, R. F., and van den Ende, H., 1997a, An experimental test for synergistic epistasis and its application in *Chlamydomonas*, *Genetics*, **145:**815–819.

De Visser, J. A. G. M., Hoekstra, R. F., and van den Ende, H., 1997b, Test of interaction between genetic markers that affect fitness in *Aspergillus niger*, *Evolution*, **51:**1499–1505.

Doerfler, W., Hoeveler, A., Weisshaar, B., Dobrzanski, P., Knebel, D., Langner, K., Achten, S., and Muller, U., 1989, Promoter Inactivation or inhibition by sequence-specific methylation and mechanisms of reactivation, *Cell Biophysics*, **15:**21–27.

Dougherty, E. C., 1955, Comparative evolution and the origin of sexuality, *Syst. Zool.*, **4:**145–190.

Drake, J. W., 1974, The role of mutation in bacterial evolution, *Symp. Soc. Gen. Microbiol.*, **24:**41–58.

Drake, J. W., 1999, The distribution of rates of spontaneous mutation over viruses, prokaryotes, and eukaryotes, *Ann. N. Y. Acad. Sci.*, **18:**100–107.

Drake, J. W., and Holland, J. J., 2000, Mutation rates among RNA viruses, *Proc. Natl. Acad. Sci. USA*, **96:**13,910–13,913.

Duarte, E., Clarke, D., Domingo, E., and Holland, J., 1992, Rapid fitness losses in mammalian RNA virus clones due to Muller's ratchet, *Proc. Natl. Acad. Sci. USA*, **89:**6015–6019.

Dubnau, D., 1991, Genetic competence in *Bacillus subtilis*, *Microbiol. Rev.*, **55:**395–424.

Dubnau, D., 1999, DNA uptake in bacteria. *Ann. Rev. Microbiol.*, **53:**217–244.

Durand, J., Birdsell, J. A., and Wills, C., 1993, Pleiotropic effects of heterozygosity at the mating-type locus of the yeast *Saccharomyces cerevisiae* on repair, recombination, and transformation, *Mut. Res.*, **290:**239–247.

Dybdahl, M. F., and Lively, C. M., 1995, Diverse, endemic and polyphyletic clones in mixed populations of the freshwater snail *Potamopyrgus antipodarum, J. Evol. Bio.*, **8:**385–398.

Dybdahl, M. F., and Lively, C. M., 1998, Host-parasite coevolution: Evidence for rare advantage and time-lagged selection in a natural population, *Evolution*, **52:**1057–1066.

Eigen, M., Gardiner, W., Schuster, P., and Winkler-Oswatitsch, R., 1981, The origin of genetic information, *Sci. Am.*, **244:**88–119.

Elena, S. F., and Lenski, R. E., 1997, Test of synergistic interactions among deleterious mutations in bacteria, *Nature*, **390:**395–398.

Eshel, I., and Feldman, M. W., 1970, On the evolutionary effects of recombination, *Theor. Pop. Bio.*, **1:**88–100.

Esposito, M. S., and Wagstaff, J. E., 1981, Mechanisms of mitotic recombination, in: *The Molecular Biology of the Yeast Saccharomyces* (J. Strathern, E. Jones, and J. Broach, eds.), pp. 341–370, Cold Spring Harbor Laboratory, Cold Spring Harbor.

Eyre-Walker, A., and Keightley, P. D., 1999, High genomic deleterious mutation rates in hominids, *Nature* **397:**344–347.

Fabre, F., and Roman, H., 1977, Genetic evidence for inducibility of recombination competence in yeast, *Proc. Natl. Acad. Sci. USA*, **74:**1667–1671.

Felkner, I. C., and Wyss, O., 1964, A substance produced by competent *Bacillus cereus 569* cells that affects transformability, *Biochem. & Biophys. Res. Comm.*, **16:**94–99.

Felsenstein, J., 1974, The evolutionary advantage of recombination, *Genetics*, **78:**737–756.

Felsenstein, J., 1988, Sex and the evolution of recombination, in: *The Evolution of Sex* (R. E. Michod and B. R. Levin, eds.), pp. 74–86, Sinauer and Associates, Sunderland.

Fisher, R. A., 1930, *The Genetical Theory of Natural Selection*, Clarendon Press, Oxford.

Fisher, R. A., 1935, The sheltering of lethals, *Am. Nat.*, **69:**446–455.

Fogel, S., and Hurst, D. D., 1963, Coincidence relations between gene conversion and mitotic recombination in *Saccharomyces*, *Genetics*, **48:**321–328.

Friedberg, E. C., 1988, Deoxyribonucleic acid repair in the yeast *Saccharomyces cerevisiae*, *Microbio. Rev.*, **52:**70–102.

Friis, J., and Roman, H., 1968, The effect of the mating-type alleles on intragenic recombination in yeast, *Genetics*, **59:**33–36.

Frischer, M. E., Thurmond, J. M., and Paul, J. H., 1990, Natural plasmid transformation in a high-frequency-of-transformation marine *Vibrio* strain, *App. Environ. Microbiol.*, **56:**3439–3444.

Fry, J. D., Keightley, P. D., Heinsohn, S. L., and Nuzhdin, S. V., 1999, New estimates of the rates and effects of mildly deleterious mutation in *Drosophila melanogaster*, *Proc. Natl. Acad. Sci. USA*, **96:**574–579.

Futuyma, D. J., 1986, *Evolutionary Biology*, Sinauer Associates, Sunderland.

Gabriel, W., Lynch, M., and Burger, R., 1993, Muller's ratchet and mutational meltdowns, *Evolution*, **47:**1744–1757.

Garcia-Dorado, A., 1997, The rate and effects distribution of viability mutation in *Drosophila:* minimum distance estimation. *Evolution*, **51:**1130–1139.

Ghiselin, M. T., 1974, *The Economy of Nature and the Evolution of Sex*, University of California Press, Berkeley.

Giannelli, F., Anagnostopoulos, T., and Green, P. M., 1999, Mutation rates in humans. II. Sporadic mutation-specific rates and rate of detrimental human mutations inferred from Hemophilia B, *Am. J. Hum. Genet.*, **65:**1580–1587.

Giannelli, F., and Green, P. M., 2000, The X chromosome and the rate of deleterious mutations in humans, *Am. J. Hum. Genet.*, **67:**515–517.

Gibson, T. C., Scheppe, M. L., and Cox, E. C., 1970, Fitness of an *Escherichia coli* mutator gene, *Science*, **169:**686–688.

Gillin, F. D., and Nossal, N. G., 1976, Control of mutation frequency by bacteriophage T4 DNA polymerase: I. The CB120 antimutator DNA polymerase is defective in strand displacement, *J. Biol. Chem.*, **251:**5219–5224.

Goldberg, I. D., Gwinn, D. D., and Thorne, C. B., 1966, Interspecies transformation between *Bacillus subtilis* and *Bacillus licheniformis*, *Biochem. & Biophys. Res. Comm.*, **23:**543–548.

Goodgal, S. H., 1982, DNA uptake in *Haemophilus* transformation, *Ann. Rev. Genet.*, **16:**169–192.

Goodman, S. D., and Scocca, J. J., 1988, Identification and arrangement of the DNA sequence recognized in specific transformation of *Neisseria gonorrhoeae*, *Proc. Natl. Acad. Sci. USA*, **85:**6982–6986.

Griffiths, A. J. F., Gelbart, W. M., Miller, J. H., and Lewontin, R. C., 1999, *Modern Genetic Analysis*, W. H. Freeman & Co., New York.

Grigorieva, G., and Shestakov, S., 1982, Transformation in the cyanobacterium *Synechosystis* sp. 6803, *FEMS Microbiol. Lett.*, **13:**367–370.

Gromkova, R., and Goodgal, S., 1979, Transformation by plasmid and chromosomal DNAs *Haemophilus parainfluenzae*, *Biochem. & Biophys. Res. Comm.*, **88:**1428–1434.

Guenther, C., 1906, *Darwinism and the Problems of Life. A Study of Familiar Animal Life* (McCabe, J., Trans.), A. Owen publishing, London.

Haas, R., Meyer, T. F., and van Putten, J. M. P., 1993, Aflagellated mutants of *Helicobacter pylori* generated by genetic transformation of naturally competent strains using transposon shuttle mutagenesis, *Mol. Microbiol.*, **8:**753–760.

Hadchouel, M., Farza, H., Simon, D., Tiollais, P., and Pourcel, C., 1987, Maternal inhibition of hepatitis B surface antigen gene expression in transgenic mice correlates with *de novo* methylation, *Nature*, **329:**454–456.

Haigh, J., 1978, The accumulation of deleterious genes in a population-Muller's ratchet, *Theor. Pop. Bio.*, **14:**251–267.

Hamilton, W. D., Axelrod, R., and Tanese, R., 1990, Sexual reproduction as an adaptation to resist parasites (a review), *Proc. Natl. Acad. Sci. USA*, **87:**3566–3573.

Hanley, K. A., Fisher, R. N., and Case, T. J., 1995, Lower mite infestations in an asexual gecko compared with its sexual ancestors, *Evolution*, **49:**418–426.

Hansen, M. T., 1978, Multiplicity of genome equivalents in the radiation-resistant bacterium *Micrococcus radiodurans*, *J. Bact.*, **134:**71–75.

Hare, J. T., and Taylor, J. H., 1989, Methylation in eucaryotes influences the repair of G/T and A/C DNA basepair mismatches, *Cell Biophys.*, **15:**29–40.

Harris, E. H., 1989, *The Chlamydomonas Sourcebook: A Comprehensive Guide to Biology and Laboratory Use*, Academic Press, San Diego.

Harsojo, S., Kitayama, S., and Matsuyama, A., 1981, Genome multiplicity and radiation resistance in *Micrococcus radiodurans*, *J. Biochem.* (*Tokoyo*) **90:**877–880.

Heller, R., and Maynard Smith, J., 1979, Does Muller's ratchet work with selfing? *Genet. Res.*, **32:**289–293.

Henaut, A., and Luzzati, M., 1972, Controle de l'aptitude a recombiner pendant la phase vegetative chez *Saccharomyces cerevisiae*, *Mol. Gen. Genet.*, **116:**26–34.

Herskowitz, I., 1988, Life cycle of the budding yeast *Saccharomyces cerevisiae*, *Microbiol. Rev.*, **52:**536–553.

Hickey, D., and Rose, M., 1988, The role of gene transfer in the evolution of eukaryotic sex, in: *The Evolution of Sex* (R. E. Michod and B. R. Levin, eds.), pp. 161–175, Sinauer Associates, Sunderland.

Hickey, D. A., 1982, Selfish DNA: A sexually-transmitted nuclear parasite, *Genetics*, **101:**519–531.

Hickey, D. A., 1993, Molecular symbionts and the evolution of sex, *J. Hered.*, **84:**410–414.

Hill, W. G., and Robertson, A., 1966, The effect of linkage on limits to artificial selection, *Genet. Res.* **8:**269–294.

Hoelzer, M. A., and Michod, R. E., 1991, DNA repair and the evolution of transformation in *Bacillus subtilis*. III. Sex with damaged DNA, *Genetics*, **128:**215–223.

Hofer, F., 1985, Transfer of lactose-fermenting ability in *Lactobacillus lactis*, *N. Z. J. Dairy Sci. Tech.*, **20:**179–183.

Holbeck, S. L., and Strathern, J. N., 1997, A role for REV3 in mutagenesis during double-strand break repari in *Saccharomyces cerevisiae*, *Genetics*, **147:**1017–1024.

Holbeck, S. L., and Strathern, J. N., 1999, *EXO1* of *Saccharomyces cerevisiae* functions in mutagenesis during double-strand break repair, *Ann. N. Y. Acad. Sci.*, **18:**375–377.

Holland, J., de la Torre, J. C., Clarke, D. K., and Duarte, E., 1991, Quantitation of relative fitness and great adaptibility of clonal populations of RNA viruses. *J. Virology*, **65:**2960–2967.

Holliday, R., 1964, A mechanism for gene conversion in fungi, *Genet. Res.*, **5:**282–304.

Holliday, R., 1984, The biological significance of meiosis, in: *Controlling Events in Meiosis* (G. D. Evans, ed.), Cambridge University Press, Cambridge.

Holliday, R., 1987, X-chromosome reactivation, *Nature*, **327:**661–662.

Holliday, R., 1988, A possible role for meiotic recombination in germline reprogramming and maintenance, in: *The Evolution of Sex* (R. E. Michod and B. R. Levin, eds.), pp. 45–55, Sinauer Associates, Sunderland.

Holliday, R., 1989a, A different kind of inheritance, *Sci. Am.*, **260:**60–73.

Holliday, R., 1989b, DNA methylation and epigenetic mechanisms, *Cell Biophys.*, **15:**15–20.

Hopf, F. A., Michod, R. E., and Sanderson, M. J., 1988, The effect of the reproductive system on mutation load, *Theor. Pop. Bio.*, **33:**243–265.

Hopwood, D. A., and Wright, H. M., 1972, Transformation in *Thermoactinomyces vulgaris*, *J. Gen. Microbiol.*, **71:**383–398.

Hoy, C. A., Fuscoe, J. C., and Thompson, L. H., 1987, Recombination and ligation of transfected DNA in CHO mutant EM9, which had high levels of sister chromatid exchange, *Mol. & Cell. Bio.*, **7:**2007–2011.

Ito, M., Fukuoda, Y., and Murata, K., 1983, Transformation of intact yeast cells treated with alkali cations, *J. Bact.*, **153:**163–168.

Jahner, D., Stuhlmann, H., Stewart, C. L., Harbers, K., Lohler, J., Simon, I., and Jaenisch, R., 1982, De novo methylation and expression of retroviral genomes during mouse embryogenesis, *Nature*, **298:**623–628.

Juni, E., 1974, Simple genetic transformation assay for rapid diagnosis of *Moraxella osloensis*, *App. Microbiol.*, **27:**16–24.

Juni, E., 1977, Genetic transformation assays for identification of strains of *Moraxella urethralis*, *J. Clin. Microbiol.*, **5:**227–235.

Juni, E., and Heym, G. A., 1980, Transformation assay for identification of psychrotrophic *Achromobacters*, *App. Environ. Microbiol.*, **40:**1106–1114.

Juni, E., Heym, G. A., and Newcomb, R. D., 1988, Identification of *Moraxella bovis* by qualitative genetic transformation and nutritional assays, *App. Environ. Microbiol.*, **54:**1304–1306.

Juni, E., and Janik, A., 1969, Transformation of *Acinetobacter calcoaceticus* (*Bacterium anitratum*), *J. Bact.*, **98:**281–288.

Kawano, S., Takano, H., Mori, K., and Kuroiwa, T., 1991, A mitochondrial plasmid that promotes mitochondrial fusion in *Physarum polycephalum*. *Protoplasma*, **160:**167–169.

Keightley, P. D., 1994, The distribution of mutation effects on viability in *Drosophila melanogaster*, *Genetics* **138:**1315–1322.

Keightley, P. D., 1998, Inference of genome-wide mutation rates and distribution of mutation effects for fitness traits: A simulation study, *Genetics*, **150:**1283–1293.

Keightley, P. D., and Bataillon, T. M., 2000, Multigeneration maximum-likelihood analysis applied to mutation-accumulation experiments in *Caenorhabditis elegans*, *Genetics*, **154:**1193–1201.

Keightley, P. D., and Caballero, A., 1997, Genomic mutation rates for lifetime reproductive output and lifespan in *Caenorhabditis elegans*, *Proc. Natl. Acad. Sci. USA*, **94:**3823–3827.

Keightley, P. D., and Eyre-Walker, A., 1999, Terumi Mukai and the riddle of deleterious mutation rates, *Genetics* **153:**515–523.

Keightley, P. D., and Eyre-Walker, A., 2000, Deleterious mutation and the evolution of sex, *Science* **290:**331–333.

Keightley, P. D., and Eyre-Walker, A., 2001, Resonse to Kondrashov, *Trends in Genetics*, **17:**77–78.

Kersulyte, D., Chalkauskas, H., and Berg, D. E., 1999, Emergence of recombinant strains of *Helicobacter pylori* during human infection, *Mol. Microbiol.* **31:**31–43.

Kibota, T. T., and Lynch, M., 1996, Estimate of the genomic mutation rate deleterious to overall fitness in *E. coli*. *Nature* **381:**694–696.

Klar, A. J. S., and Bonaduce, M. J., 1993, The mechanism of fission yeast mating-type interconversioni: evidence for two types of epigenetically inherited chromosomal imprinting events, *Cold Spring Harbor Symp. Quant. Bio.*, **58:**457–465.

Kondrashov, A. S., 1982, Selection against harmful mutations in large sexual and asexual populations, *Genet. Res.*, **40:**325–332.

Kondrashov, A. S., 1988, Deleterious mutations and the evolution of sexual reproduction, *Nature.*, **336:**435–440.

Kondrashov, A. S., 1993, Classification of hypotheses on the advantage of amphimixis, *J. Hered.*, **84:**372–387.

Kondrashov, A. S., 1994c, Muller's ratchet under epistatic selection, *Genetics*, **136:**1469–1473.

Kondrashov, A. S., 1998, Measuring spontaneous deleterious mutation process, *Genetica*, **102/103:**183–197.

Kondrashov, A. S., 2001, Sex and U, *Trends in Genetics*, **17:**75–77.

Kowalski, S., and Laskowski, W., 1975, The effect of three *rad* genes on survival, inter- and intragenic mitotic recombination in *Saccharomyces*, *Mol. & Gen. Genet.*, **136:**75–86.

Koyama, Y., Hoshino, T., Tomizuka, N., and Furukawa, K., 1986, Genetic transformation of the extreme thermophile *Thermus thermophilus* and of other *Thermus spp*, *J. Bact.*, **166:**338–340.

Kruger, D. H., Schroeder, C., Santibanez-Koref, M., and Reuter, M., 1989, Avoidance of DNA methylation: a virus-encoded methylase inhibitor and evidence for counterselection of methylation recognition sites in viral genomes, *Cell Biophys.*, **15:**87–95.

Kunz, B. A., Barclay, B. J., Game, J. C., Little, J. G., and Haynes, R. H., 1980, Induction of mitotic recombination in yeast by starvation for thymine nucleotides, *Proc. Natl. Acad. Sci. USA*, **77:**6057–6061.

Kunz, B. A., and Haynes, R. H., 1981a, DNA repair and mutagenesis in yeast, in: *The Molecular Biology of the Yeast Saccharomyces* (J. Strathern, E. Jones, and J. Broach, eds.), pp. 371–414, Cold Spring Harbor Laboratory, Cold Spring Harbor.

Kunz, B. A., and Haynes, R. H. 1981b, Phenomenology and genetic control of mitotic recombination in yeast. *Ann. Rev. Genet.*, **15:**57–89.

Lambert, J. D., and Moran, N. A., 1998, Deleterious mutations destabilize ribosomal RNA in endosymbiotic bacteria, *Proc. Natl. Acad. Sci. USA*, **95:**4458–4462.

Lang, B. F., Burger, G., O'Kelly, C. J., Cedergren, R., Golding, G. B., Lemieux, C., Sankoff, D., Turmel, M., and Gray, M. W., 1997, An ancestral mitochondrial DNA resembling a eubacterial genome in miniature, *Nature*, **387:**493–497.

Lemontt, J. F., 1980, Genetic and physiological factors affecting repair and mutagenesis in yeast, in: *DNA repair and mutagenesis in eukaryotes* (F. J. de Serres, ed.), pp. 85–120, Plenum Press, New York.

Li, E., Beard, C., and Jaenisch, R., 1993, Role for DNA methylation in genomic imprinting, *Nature*, **366:**362–365.

Linz, B., Schenker, M., Zhu, P., and Achtman, M., 2000, Frequent interspecific genetic exchange between commensal neisseriae and *Neisseria meningitidis*, *Mol. Microbiol.*, **36:**1049–1058.

Lively, C., Craddock, C., and Vrijenhoek, R., 1990, Red queen hypothesis supported by parasitism in sexual and clonal fish, *Nature*, **344:**864–866.

Lorenz, M. G., Gerjets, D., and Wackernagel, W., 1991, Release of transforming plasmid and chromosomal DNA from two cultured soil bacteria, *Arch. Microbiol.*, **156:**319–326.

Lorenz, M. G., and Wackernagel, W., 1993, Transformation as a mechanism for bacterial gene transfer in soil and sediment—studies with a sand/clay microcosm and the cyanobacterium *Synechocystis* OL50, in: *Trends In Microbial Ecology* (R. Guerrero and C. Pedros-Alio, eds.), pp. 325–330, Spanish Society For Microbiology, Barcelona.

Lorenz, M. G., and Wackernagel, W., 1994, Bacterial gene transfer by natural genetic transformation in the environment, *Microbiol. Rev.*, **58:**563–602.

Lynch, M., and Blanchard, J. L., 1998, Deleterious mutation accumulation in organelle genomes, *Genetica*, **102/103:**29–39.

Lynch, M., Burger, R., Butcher, D., and Gabriel, W., 1993, The mutational meltdown in asexual populations, *J. Hered.*, **84:**339–344.

Lynch, M., and Gabriel, W., 1990, Mutation load and the survival of small populations, *Evolution*, **44:**1725–1737.

Kibota, T., and Lynch, M., 1996, Estimate of the genomic mutation rate deleterious to overall fitness in *E. coli*, *Nature*, **381:**694–696.

Magni, G. E., 1963, The origin of spontaneous mutations during meiosis, *Proc. Natl. Acad. Sci. USA*, **50:**975–980.

Magni, G. E., and von Borstel, R. C., 1962, Different rates of spontaneous mutation during mitosis and meiosis in yeast, *Genetics* **47:**1097–1108.

Margulis, L., 1981, *Symbiosis in Cell Evolution*, W. H. Freeman and Company, San Francisco.

Margulis, L., and Sagan, D., 1984, Evolutionary origins of sex, in: *Oxford Surveys in Evolutionary Biology* (R. Dawkins and M. Ridley, eds.), pp. 16–47, Oxford University Press, London.

Margulis, L., and Sagan, D., 1985, *Origins of Sex*, Yale University Press, New Haven.

Margulis, L., Sagan, D., and Olendzenski, L., 1985, What is sex? in: *The Origin and Evolution of Sex* (H. O. Halvorson and A. Monroy, eds.), pp. 69–85, Marine Biological Laboratory, Woods Hole, Alan R. Liss Inc., New York.

Mathis, L. S., and Scocca, J. J., 1982, *Haemophilus influenzae* and *Neisseria gonorrhoeae* recognize different specificity determinants in the DNA uptake step of genetic transformation, *J. Gen. Microbiol.*, **128:**1159–1161.

Mattimore, V., and Battista, J. R., 1996, Radioresistance of *Deinococcus ragiodurans*: Functions necesary to survive ionizing radiation are also necessary to survive prolonged desiccation, *J. Bact.*, **178:**633–637.

Matzke, M., and Matzke, A. J. M., 1993, Genomic imprinting in plants: parental effects and trans-inactivation phenomena, *Ann. Rev. Plant Phys. & Plant Mol. Bio.*, **44:**53–76.

Maynard Smith, J., 1968, Evolution in sexual and asexual populations, *Am. Nat.* **102:**469–473.

Maynard Smith, J., 1971, What use is sex? *J. Theor. Biol.*, **30:**319–335.

Maynard Smith, J., 1978, *The Evolution of Sex*, Canbridge University Press, Cambridge.

Maynard Smith, J., 1988, The evolution of recombination, in: *The Evolution of Sex* (R. E. Michod and B. R. Levin, eds.) pp. 106–125, Sinauer and Associates, Sunderland.

Maynard Smith, J., and Smith, N. H., 1998, Detecting recombination from gene trees, *Mol. Biol. Evol.*, **15:**590–599.

McClain, M. E., and Spendlove, R. S., 1966, Multiplicity reactivation of reovirus particles after exposure to ultraviolet light, *J. Bact.*, **92:**1422–1429.

McGill, C. B., Holbeck, S. L., and Strathern, J. N., 1998, The chromosomal bias of misincorporation during double-strand break repair is not altered in mismatch repair-defective strains of *Saccharomyces cerevisiae*, *Genetics* **148:**1525–1533.

McGrath, J., and Solter, D., 1984, Completion of mouse embryogenesis requires both the maternal and paternal genomes, *Cell*, **37:**179–183.

Melzer, A. L., and Koeslag, J. H., 1991, Mutations do not accumulate in asexual isolates capable of growth and extinction-Muller's ratchet re-examined, *Evolution*, **45:**649–655.

Mereschkovsky, C., 1905, Le plante consideree comme une complex symbiotique, *Bulletin Societe Science Naturelle, Ouest*, **6:**17–98.

Metzenberg, R. L., and Glass, N. L., 1990, Mating type and mating strategies in *Neurospora*, *BioEssays* **12:**53–59.

Mevarech, M., and Werczberger, R., 1985, Genetic transfer in *Halobacterium volcanii*, *J. Bact.*, **162:**461–462.

Michod, R., 1993, Genetic error, sex, and diploidy, *J. Hered.*, **84:**360–371.

Michod, R. E., 1990, Evolution of sex, *Trends Ecol. Evol.* **5:**30.

Michod, R. E., and Gayley, T. W., 1992, Masking of mutations and the evolution of sex, *Am. Nat.*, **139:**706–734.

Michod, R. E., and Levin, B. R., 1988, Introduction, in: *The Evolution of Sex* (R. E. Michod and B. R. Levin, eds.), pp. 1–6, Sinauer Associates, Sunderland.

Michod, R. E., and Long, A., 1995, Origin of sex for error repair. II. Rarity and extreme environments. *Theor. Pop. Biol.* **47:**56–81.

Michod, R. E., Wojciechowski, M. F., and Hoelzer, M. A., 1988, DNA repair and the evolution of transformation in the bacterium *Bacillus subtilis*, *Genetics*, **118:**31–39.

Miller, S. L., and Orgel, L. E., 1974, *The Origins of Life on the Earh* (W. D. McElroy and C. P. Swanson, eds.), Prentice-Hall, Englewood Cliffs.

Mongold, J. A., 1992, DNA repair and the evolution of transformation in *Haemophilus influenzae*, *Genetics* **132:**893–898.

Monk, M., 1986, Methylation and the X chromosome, *BioEssays*, **4:**204–208.

Monk, M., 1987, Memories of mother and father, *Nature*, **328:**203–204.

Moore, T., and Haig, D., 1991, Genomic imprinting in mammalian development: a parental tug-of-war, *Trends Genet.*, **7:**45–49.

Moran, N. A., 1996, Accelerated evolution and Muller's ratchet in endosymbiotic bacteria, *Proc. Natl. Acad. Sci. USA*, **96:**2873–2878.

Morgan, T. H., 1913, *Heredity and Sex*, Columbia University Press, New York.
Moritz, C., McCallum, H., Donnellan, S., and Roberts, J. D., 1991, Parasite loads in parthenogenetic and sexual lizards (*Heteronotia binoei*): support for the Red Queen hypothesis, *Proc. R. Soc. Lond., B*, **244:**145–149.
Morrison, D. A., Mannarelli, B., and Vijayakumar, M. N., 1982, Competence for transformation in *Streptococcus pneumoniae*: an inducible high-capacity system for genetic exchange, in: *Microbiology* (D. Schlessinger, ed.), *Am. Soc. Microbiol.*, Washington, D. C.
Mukai, T., 1964, The genetic structure of natural populations of *Drosophila melanogaster*. I. Spontaneous mutation rate of polygenes controlling viability, *Genetics*, **50:**1–19.
Mukai, T., and Yamazaki, T., 1964, Position effect of spontaneous mutant polygenes controlling viability in *Drosophila melanogaster*, *Proc. Japan Acad.*, **40:**840–845.
Mukai, T., 1969, The genetic structure of natural populations of *Drosophila melanogaster*. VII Synergistic interaction of spontaneous mutant polygenes controlling viability, *Genetics*, **61:**749–761.
Mukai, T., Chigusa, S. T., Mettler, L. E., and Crow, J. F., 1972, Mutation rate and dominance of genes affecting viability in *Drosophila melanogaster*, *Genetics*, **72:**335–355.
Muller, H. J., 1914, A gene for the fourth chromosome of *Drosophila*, *J. Exp. Zoo.*, **17:**325–336.
Muller, H. J., 1918, Genetic variability, twin hybrids and constant hybrids, in a case of balanced lethal factors, *Genetics*, **3:**422–500.
Muller, H. J., 1932, Some genetic aspects of sex, *Am. Nat.*, **66:**118–138.
Muller, H. J., 1958, Evolution by mutation, *Bull. Am. Math. Soc.*, **64:**137–160.
Muller, H. J., 1964, The relation of recombination to mutational advance, *Mut. Res.*, **1:**2–9.
Nachman, M. W., and Crowell, S. L., 2000, Estimate of the mutation rate per nucleotide in humans, *Genetics*, **156:**297–304.
Nei, M., 1970, Accumulation of nonfunctional genes on sheltered chromosomes, *Am. Nat.*, **104:**311–322.
Nevoigt, E., Fassbender, A., and Stahl, U., 2000, Cells of the yeast *Saccharomyces cerevisiae* are transformable by DNA under non-artificial conditions, *Yeast* **16:**1107–1110.
Niwa, O., and Sugahara, T., 1981, 5-Azacytidine induction of mouse endogenous type C virus and suppression of DNA methylation, *Proc. Natl. Acad. Sci. USA*, **78:**6290–6294.
Norgard, M. V., and Imaeda, T., 1978, Physiological factors involved in the transformation of *Mycobacterium smegmatis*, *J. Bact.*, **133:**1254–1262.
Nur, I., Pascale, E., and Furano, A. V., 1989, Demethylation and specific remethylation of the promoter-like reion of the L family of mammalian transposable elements, *Cell Biophys.*, **15:**61–66.
O'Conner, M., Wopat, A., and Hanson, R. S., 1977, Genetic transformation in *Methylobacterium organophilum*, *J. Gen. Microbiol.*, **98:**265–272.
Ohnishi, O., 1977, Spontaneous and ethyl methanesulfonate-induced mutations controlling viability in *Drosophila melanogaster.* II. Homozygous effect of polygenetic mutations, *Genetics*, **87:**529–545.
Okimoto, R., Macfarlane, J. L., Clary, D. O., and Wolstenholme, D. R., 1992, The mitochondrial genomes of two nematodes, *Caenorhabditis eloegans* and *Ascaris suum*, *Genetics*, **130:**471–498.
Ormerod, J. G., 1988, Natural genetic transformation in *Chlorobium*, in: *Green Photosynthetic Bacteria* (J. M. Olson, J. Ormerod, J. Amesz, E. Stackebrand T, and H. G. Truper, eds.), pp. 315–319, Plenum Press, New York.
Otto, S. P., and Feldman, M. W., 1997, Deleterious mutations, variable epistatic interactions, and the evolution of recombination, *Theor. Pop. Bio.*, **51:**134–147.
Ottolenghi, E., and Hotchkiss, R. D., 1960, Appearance of genetic transforming activity in pneumococcal cultures, *Science*, **132:**1257–1258.

Page, W. J., 1981, Optimal conditions for induction of competence in nitrogen-fixing *Azotobacter vinelanii*, Can. J. Microbiol., **28:**389–397.

Pakula, R., and Walczak, W., 1963, On the nature of competence of transformable Streptococci, *J. Gen. Microbiol.*, **31:**125–133.

Pickett-Heaps, J. D., 1971, The autonomy of the centriole: fact or fallacy? *Cytobios*, **3:**205–214.

Pifer, M. L., and Smith, H. O., 1985, Processing of donor DNA during *Haemophilus influenzae* transformation: analysis using a model plasmid system, *Proc. Natl. Acad. Sci. USA*, **82:**3731–3735.

Redfield, R. J., 1988, Evolution of bacterial transformation: Is sex with dead cells ever better than no sex at all? *Genetics*, **119:**213–221.

Redfield, R. J., 1993a, Evolution of transformation: testing the DNA repair hypothesis in *Bacillus subtilis* and *Haemophilus influenzae*, *Genetics*, **133:**755–761.

Redfield, R. J., 1993b, Genes for breakfast: the have-your-cake-and -eat-it-too of bacterial transformation, *J. Hered.*, **84:**400–404.

Redfield, R. J., Schrag, M. R., and Dean, A. M., 1997, The evolution of bacterial transformation: Sex with bad relations, *Genetics*, **146:**27–38.

Reik, W., Collick, A., Norris, M. L., Barton, S. C., and Surani, M. A., 1987, Genomic imprinting determines methylation of parental alleles in transgenic mice, *Nature*, **328:**248–254.

Resnick, M. A., 1976, The repair of double-strand breaks in DNA: A model involving recombination, *J. Theor. Bio.*, **59:**97–106.

Reynolds, R., 1987, Induction and repair of closely opposed pyrimidine dimers in *Saccharomyces cerevisiae*, *Mut. Res.*, **184:**197–207.

Riggs, A. D., 1989, DNA methylation and cell memory, *Cell Biophys.*, **15:**1–13.

Rodarte-Ramon, U. S., 1972, Radiation induced recombination in *Saccharomyces*: the genetic control of recombination in mitosis and meiosis, *Rad. Res.*, **49:**148–154.

Rodarte-Ramone, U. S., and Mortimer, R. K., 1972, Radiation induced recombination in *Saccharomyces*: isolation and genetic study of recombination deficient mutants, *Rad. Res.*, **49:**133–147.

Roelants, P., Konvalinkova, V., Mergeay, M., and Lurquin, P. F., 1976, DNA uptake by *Streptomyces* species, *Biochim. Biophys. Acta*, **442:**117–122.

Rosche, W. A., and Foster, P. L., 2000, Determining mutation rates in bacterial populations, *Methods* **20:**4–17.

Rose, M. R., 1983, The contagion mechanism for the origin of sex, *J. Theor. Bio.*, **101:**137–146.

Rowlands, R. T., and Turner, G., 1974, Recombination between the extranuclear genes conferring oligomycin resistance and cold sensitivity in *Aspergillus nidulans*, *Mol. Gen. Genet.*, **133:**151–161.

Rudin, I., Sjostrom, J. E., Lindberg, M., and Philipson, L., 1974, Factors affecting competence for transformation in *Staphylococcus aureus*, *J. Bact.*, **118:**155–164.

Russell, P. J., 1998, *Genetics*, Benjamin/Cummings, Menlo Park.

Sagan, C., 1973, Ultraviolet selection pressures on the earliest organisms, *J. Theor. Bio.*, **39:**195–200.

Sager, R., and Granick, S., 1954, Nutritional control of sexuality in *Chlamydomonas reinhardi*, *J. Gen. Physiol.* **37:**729–742.

Schultz, S. T., Lynch, M., and Willis, J. H., 1999, Spontaneous deleterious mutation in *Arabidopsis thaliana*, *Proc. Natl. Acad. Sci. USA*, **96:**393–11398.

Seger, J., and Hamilton, W. D., 1988, Parasites and sex, in: *The Evolution of Sex* (R. E. Michod and B. R. Levin, eds.), pp. 176–193, Sinauer and Associates, Sunderland.

Selk, E., and Wills, C., 1997, Mismatch repair and the accumulation of deleterious mutations influence the competitive advantage of *MAT* (mating type) heterozygosity in the yeast *Saccharomyces cerevisiae*, *Genet. Res.* **71:**1–10.

Shabalina, S. A., Yampolsky, L. Y., and Kondrashov, A. S., 1997, Rapid decline of fitness in panmictic populations of *Drosophila melanogaster* maintained under relaxed natural selection, *Proc. Natl. Acad. Sci. USA*, **94:**13034–13039.

Shah, G. R., and Caufield, P. W., 1993, Enhanced transformation of *Streptococcus mutans* by modifications in culture conditions, *Anal. Biochem.*, **214:**343–346.

Shaw, R. G., Byers, D. L., and Darmo, E., 2000, Spontaneous mutational effects on reproductive traits of *Arabidopsis thaliana*, *Genetics*, **155:**369–378.

Shestakov, S. V., and Khyen, N. T., 1970, Evidence for genetic transformation in blue-green alga *Anacystis nidulans*, Mol. Gen. Genet., **107:**372–375.

Simmons, M. J., and Crow, J. F., 1977, Mutations affecting fitness in *Drosophila* populations, *Ann. Re. Genet.*, **11:**49–78.

Sinha, R. P., and Iyer, V. N., 1971, Competence for genetic transformation and the release of DNA from *Bacillus subtilis*, *Biochim. Biophys. Acta*, **232:**61–71.

Sisco, K. L., and Smith, H. O., 1979, Sequence-specific DNA uptake in *Haemophilus* transformation, *Proc. Natl. Acad. Sci. USA*, **76:**972–976.

Smith, H. O., Gwinn, M. L., and Salzberg, S. L., 1999, DNA uptake signal sequences in naturally transformable bacteria, *Res. Microbiol.*, **150:**603–616.

Sniegowski, P. D., Gerrish, P. J., Johnson, T., and Shaver, A., 2000, The evolution of mutation rates: separating causes from consequences, *BioEssays*, **22:**1057–1066.

Sniegowski, P. D., Gerrish, P. J., and Lenski, R. E., 1997, Evolution of high mutation rates in experimental populations of *E. coli*, *Nature*, **387:**703–705.

Snustad, D. P., Simmons, M. J., and Jenkins, J. B., 1997, *Principles of Genetics*, John Wiley & Sons, New York.

Sparling, P. F., 1966, Genetic transformation of *Neisseria gonorrhoeae* to streptomycin resistance, *J. Bact.*, **92:**1364–1371.

Stearns, S. C., 1987, *The Evolution of Sex and its Consequences*, Sinauer Associates, Sunderland.

Stevens, S. E., and Porter, R. D., 1986, Heterospecific transformation among cyanobacteria, *J. Bact.*, **167:**1074–1-76.

Stewart, G., and Carlson, C. A., 1986, The biology of natural transformation, *Ann. Rev. Microbiol.*, **40:**211–235.

Strathern, J. N., Shafer, B. K., and McGill, C. B., 1995, DNA synthesis errors associated with double-strand-break repair, *Genetics*, **140:**965–972.

Streips, U. N., and Young, F. E., 1971, Mode of action of the competence-inducing factor of *Bacillus stearothermophilus*, *J. Bact.*, **106:**868–875.

Strobeck, C., Maynard Smith, J., and Charlesworth, B., 1976, The effects of hitchhiking on a gene for recombination, *Genetics*, **82:**547–558.

Stroun, M., Anker, P., and Auderset, G., 1970, Natural release of nucleic acids from bacteria into plant cells, *Nature*, **227:**607–608.

Stryer, L., 1988, *Biochemistry* (3rd ed.), W. H. Freeman, New York.

Sturtevant, A. H., and Mather, K., 1938, The interrelations of inversions, heterosis, and recombination, *Am. Nat.*, **72:**447–452.

Suerbaum, S., and Achtman, M., 1999, Evolution of *Helicobacter pylori*: the role of recombination, *Trends in Microbiology*, **7:**182.

Suerbaum, S., Maynard Smith, J., Bapumia, K., Morelli, G., Smith, N. H., Kunstmann, E., Dyrek, I., and Achtman, M., 1998, Free recombination within *Helicobacter pylori*, *Proc. Natl. Acad. Sci., USA*, **95:**12619–12624.

Sun, H., Treco, D., Schultes, N. P., and Szostak, J. W., 1989, Double-stranded breaks at an initiation site for meiotic gene conversion, *Nature*, **338:**87–90.

Surani, M. A., Reik, W., and Allen, N. D., 1988, Transgenes as molecular probes for genomic imprinting, *Trends Genet.*, **4:**59–62.

Surani, M. A. H., Barton, S. C., and Norris, M. L., 1984, Development of reconstituted mouse eggs suggests imprinting of the genome during gametogenesis, *Nature*, **308:**548–550.

Surani, M. A. H., Barton, S. C., and Norris, M. L., 1987, Influence of parental chromosomes on spatial specificity in androgenetic→ parthenogenetic chimeras in the mouse, *Nature*, **326:**395–397.

Swain, J. L., Stewart, T. A., and Leder, P., 1987, Parental legacy determines methylation and expression of an autosomal transgene: A molecular mechanism for parental inprinting, *Cell*, **50:**719–727.

Szostak, J. W., Orr-Weaver, T. L., Rothstein, R. J., and Stahl, F. W., 1983, The double-strand break repair model for recombination, *Cell*, **33:**25–35.

Takahashi, I., and Gibbons, N. E., 1957, Effect of salt concentration on the extracellular nucleic acids of *Micrococcus halodenitrificans*, *Can. J. Microbiol.*, **3:**687–694.

Thacker, J., 1989, The use of integrating DNA vectors to analyse the molecular defects in ionising radiation-sensitive mutants of mammalian cells including ataxia telangiectasia, *Mut. Res.*, **220:**187–204.

Thompson, L. H., Brookman, K. W., Dillehay, L. E., Carrano, A. V., Mazrimas, J. A., Mooney, C. L., and Minkler, J. L., 1982, A CHO-cell strain having hypersensitivity to mutagens, a defect in DNA strand -break repair, and an extraordinary baseline frequency of sister-chromatid exchange, *Mut. Res.* **95:**427–440.

Tilman, D., Wedin, D., and Knops, J., 1996, Productivity and sustainability influenced by biodiversity in grassland ecosystems, *Nature*, **379:**718–720.

Tirgari, S., and Moseley, B. E. B., 1980, Transformation in *Micrococcus radiodurans:* measurement of various parameters and evidence for multiple, independently segregating genomes per cell, *J. Gen. Microbiol.*, **119:**287–296.

Tomasz, A., 1965, Control of the competent state in *Pneumococcus* by a hormone like cell product: an example for a new type of regulatory mechanism in bacteria, *Nature*, **208:** 155–159.

Tomasz, A., 1966, Model for the mechanism controlling the expression of competent state in *Pneumococcus* cultures, *J. Bact.*, **91:**1050–1061.

Tomasz, A., and Hotchkiss, R. D., 1964, Regulation of the transformability of *Pneumococcal* cultures by macromolecular cell products, *Proc. Natl. Acad. Sci. USA*, **51:**480–487.

Trehan, K., and Sinha, U., 1981, Genetic transfer in a nitrogen-fixing filamentous cyanobacterium, *J. Gen. Microbiol.* **124:**349–352.

Tremblay, K. D., Saam, J. R., Ingram, R. S., Tilghman, S. M., and Bartolomei, M. S., 1995, A paternal-specific methylation imprint marks the alleles of the mouse *H19* gene, *Nat. Genet.*, **9:**407–413.

Van Nieuwenhoven, W. H., Hellingwerf, K. J., Venema, G., and Konings, W. N., 1982, Role of proton motive force in genetic transformation of *Bacillus subtilis*, *J. Bact.*, **151:**771–776.

Van Valen, L., 1973, A new evolutionary law, *Evol. Theory*, **1:**1–30.

Vassilieva, L. L., and Lynch, M., 1999, The rate of spontaneous mutation for life-history traits in *Caenorhabditis elegans*, *Genetics*, **151:**119–129.

Venema, G., 1979, Bacterial transformation, *Adv. Microb. Phys.*, **19:**245–331.

Wagner, G. P., and Gabriel, W., 1990, Quantitative variation in finite parthenogenetic populations: What stops Muller's ratchet in the absence of recombination? *Evolution*, **44:**715–731.

Wake, C. T., Vernaleone, F., and Wilson, J. H., 1985, Topological requirements for homologous recombination among DNA molecules transfected into mammalian cells, *Mol. & Cell. Biol.*, **5:**2080–2089.

Walker, I., 1978, The evolution of sexual reproduction as a repair mechanism. Part 1. A model for self-repair and its biological implications, *Acta Bio.*, **27:**133–158.

Wang, Y., and Taylor, D. E., 1990, Natural transformation in *Campylobacter* species, *J. Bact.*, **172:**949–955.

Watson, J. D., Hopkins, N. H., Roberts, J. W., Steitz, J. A., and Weiner, A. M., 1987, *Molecular Biology of the Gene* (4th ed.), Benjamin Cummings, Menlo Park.

Weismann, A., 1889, *Essays upon Heredity and Kindred Biological Problems* (E. B. Poulton, S. Schonland, and A. E. Shipley Trans.). Clarendon Press, Oxford.

West, S. A., Peters, A. D., and Barton, N. H., 1998, Testing for epistasis between deleterious mutations, *Genetics*, **149:**435–444.

Whitehouse, H. L. K., 1982, *Genetic Recombination: Understanding The Mechanisms.* Wiley, New York.

Williams, G. C., 1975, *Sex and Evolution*, Princeton University Press, Princeton.

Williams, G. C., and Mitton, J. B., 1973, Why reproduce sexually? *J. Theor. Biol.*, **39:**545–554.

Willis, J. H., 1993, Effects of different levels of inbreeding on fitness components in *Mimulus guttatus*, *Evolution*, **47:**864–876.

Wilson, V. L., and Jones, P. A., 1983, DNA methylation decreases in aging but not in immortal cells, *Science*, **220:**1055–1057.

Wojciechowski, M. F., Hoelzer, M. A., and Michod, R. E., 1989, DNA repair and the evolution of transformation in Bacillus subtilis. II. Role of inducible repair, *Genetics*, **121:**411–422.

Worrell, V. E., Nagle, D. P., McCarthy, D., and Eisenbraun, A., 1988, Genetic transformation system in the archaebacterium *Methanobacterium thermoautotrophicum* Marburg, *J. Bact.*, **170:**653–656.

Wright, S., 1932, The roles of mutation, inbreeding, crossbreeding, and selection in evolution, *Proc. XI International Congr. Genet.*, **1:**356–366.

Wright, S., 1988, Surfaces of selective value revisited, *Am. Nat.*, **131:**115–123.

Yankofsky, S. A., Gurevich, R., Grimland, N., and Stark, A. A., 1983, Genetic transformation of obligately chemolithotrophic thiobacilli, *J. Bact.*, **153:**652–657.

Zuccotti, M., and Monk, M., 1995, Methylation of the mouse *Xist* gene in sperm and eggs correlates with imprinted *Xist* expression and paternal X-inactivation, *Nat. Genet.*, **9:**316–320.

Zeyl, C., Bell, G., and Green, D. M., 1996, Sex and the spread of retrotransposon *Ty3* in experimental populations of *Saccharomyces cerevisiae*, *Genetics* **143:**1567–1577.

Zeyl, C., and DeVisser, J. A. G. M., 2000, Estimates of the rate and distribution of fitness effects of spontaneous mutation in *Saccharomyces cerevisiae*, Manuscript Submitted to Genetics.

3

Phylogenetic, Geographical, and Temporal Analysis of Female Reproductive Trade-Offs in Drosophilidae

WILLIAM T. STARMER*, MICHAL POLAK, SCOTT PITNICK, SHANE F. McEVEY, J. STUART F. BARKER, and LARRY L. WOLF

The fact that reproductive effort often shows trade-offs with other necessary functions and features of living organisms has been recognized for centuries. Darwin (1872, pg. 142) gives credit to Geoffroy St. Hilaire and Goethe for proposing the law of "Compensation or Balancement of Growth" and ascribes the following quote to Goethe, "In order to spend on one side, nature is forced to economize on the other side." The essence of this law is captured in modern theories and syntheses (Lack, 1947; Cody, 1966; Smith and Fretwell, 1974; Stearns, 1976, 1977, 1992) that emphasize

WILLIAM T. STARMER, SCOTT PITNICK, and LARRY L. WOLF • Department of Biology, 108 College Place, Syracuse University, Syracuse, NY 13244-1270 USA. MICHAL POLAK • Department of Biological Sciences, University of Cincinnati, Cincinnati, OH 45221-0006 USA. SHANE F. McEVEY • Australian Museum, 6 College Street, Sydney, NSW 2010 Australia. J. STUART F. BARKER • Department of Animal Science, University of New England, Armidale, NSW 2351 Australia.

* Author for correspondence: Email: *wstarmer@syr.edu*, Phone: (315) 443-2154, Fax: (315) 443-2156

time, energy budgets, and physiological, genetic and phylogenetic associations that govern the form of compensation that results in a trade-off.

One trade-off that is expected to be widespread is between offspring size and number. Particularly in species with no parental care, it is presumed that offspring viability increases with egg size (e.g., Smith and Fretwell, 1974; Parker and Begon, 1986; McGinley et al., 1987; Winkler and Wallin, 1987). Because energy allocated to reproduction must be divided among offspring, any increase in maternal resources devoted to each offspring should diminish the number of offspring produced. Indeed, a phenotypic trade-off between egg size and fecundity, indicative of such an underlying physiological trade-off, has been widely identified (e.g., Darwin, 1872; Lack, 1947; Stearns, 1992). To the extent that population-level responses to selection on genetic variation in this physiological trade-off occur, a microevolutionary (i.e., genetic) trade-off between egg size and number is expected. Finally, if conditions relevant to this trade-off are common to whole lineages, then a macroevolutionary trade-off between these traits may result.

In the genus *Drosophila*, the relationship between egg size and fecundity is unclear. Studies of three species (*D. simulans*, *D. subobscura* and *D. phalerata*) found no evidence of intraspecific phenotypic trade-offs between egg volume and egg number (Avelar and Rocha Pité, 1989; Avelar, 1993). An experimental study imposing bidirectional selection on egg size in *D. melanogaster* found a phenotypic correlation between egg size and fecundity in the lines selected for increased egg size, but not in the control or decreased egg size lines. No genetic correlation between egg size and fecundity was observed (Schwarzkopf et al., 1999). In contrast, significant negative phenotypic relationships between egg size and fecundity have been reported among populations of *D. hibisci* (Starmer et al., 1997; Wolf et al., 2000) and among species of Hawaiian *Drosophila* (Montague et al., 1981; Berrigan, 1991).

The expected negative relation between these traits can be expressed as a power function such that egg size = $\alpha \times (\text{egg number})^{\beta}$. The exponent β is expected to be equal to -1 when the trade-off is isometric, e.g. egg size $\times$ egg number is constant (α). In the Hawaiian drosophilids, which have egg sizes ranging from 0.014 to 0.269 mm^3 and clutch sizes from 2 to 150 (Kambysellis and Heed, 1971), β was estimated to be close to -1 depending on assumptions about ovariole activity responsible for egg number data (Montague et al., 1981). The extraordinary situation responsible for the evolution of the Hawaiian fauna (Carson and Kaneshiro, 1976) and the extreme isolation of this radiation, however, make the relationship among egg size, egg number and reproductive effort dynamics somewhat singular. It is thus desirable to compare the Hawaiian example to other drosophilid radiations of similar magnitude.

To better elucidate this important trade-off in Drosophilidae, we here determine the relationship between egg size and fecundity at multiple levels: temporal, geographic, interspecific, intergeneric, and among three discrete continental radiations. We use field data from the flower-breeding *Scaptodrosophila hibisci* (*Drosophila hibisci*) collected over a wide geographic range and over time at the same location in eastern Australia (Starmer et al., 1997, 1998, 2000; Wolf et al., 2000). Laboratory isofemale lines from this species collected at separate localities (Starmer et al., 2000) provide additional information on the intrapopulation level of variation for the trade-off and for reproductive effort. In addition, specific comparisons of field-collected versus laboratory-reared females from the same site, provide information about the laboratory effects on reproductive activity. Recently, *Scaptodrosophila aclinata*, a species closely related to *S. hibisci*, was discovered in the Northern Territory (McEvey and Barker, 2001). Data from this species are included to provide a comparison between two closely related species. Intergeneric comparisons are also examined. Finally, we compare the egg size-egg number relationship among three discrete radiations.

In addition to the Drosophilidae of Hawaii, Carson and Okada (1982), Bock (1976) and Parsons and Bock (1979) have implicated the *Scaptodrosophila* of New Guinea and Australia as another example of a large adaptive radiation. Some North American species of *Drosophila* have been investigated in conjunction with the earlier Hawaiian work (Berrigan, 1991) and a study of domestic drosophilids (Atkinson, 1979) in Europe has been reported, but comparisons between the continental and island groups have not been made. The Drosophilidae species in these three geographic regions (Hawaiian, Australasian and North American) differ in several respects. The salient differences are: 1) the geographic setting is largest for the continental North American fauna, intermediate for the Australasian and smallest for the Hawaiian; 2) the adult body size variation is much larger for the Hawaiian flies; 3) the time scale of evolution for the Hawaiian forms is shorter; 4) the fauna in Hawaii is representative of a single radiation, whereas the North American taxa include Nearctic and Neotropical components; and 5) although not entirely known, the habitats of the Hawaiian species are more varied. All of these differences can be used to argue that the functional relationship between egg number and egg size may not be the same for the three regions. Indeed, movement away from an isometric relationship and changes in reproductive effort can implicate both adaptive and non-adaptive constraints such as phylogenetics, demands on locomotor performance (Berrigan, 1991), minimal egg size (Wiklund et al., 1987), and allometric growth differences due to environmental differences.

MATERIALS AND METHODS

Data for Hawaiian taxa were taken from Kambysellis and Heed (1971). The Australasian data were collected from wild-caught females that were attracted to baits, found on natural substrates or swept from foliage and litter. These collections were made in four regions (Tasmania, central New South Wales, northern Queensland and southern New Caledonia) during January and February, 2000. Records for *S. hibisci* and *S. aclinata* were from extensive collections over the species ranges prior to the 2000 collection (Starmer et al., 1997, 1998, 2000, and Wolf et al., 2000). The North American data were mainly gathered from laboratory stock cultures obtained from the Bowling Green stock center or from our stocks. Five species (*Drosophila neotestacea, D. putrida, Hirtodrosophila duncani, Mycodrosophila claytonae, M. dimidiata*) were collected from wild mushrooms in central New York state in the summer of 2000. Female size (thorax length = **tl**), ovariole number (**ova**) and egg dimensions (egg width = **ew**, egg length = **el**) were determined by the methods described in Starmer et al. (1997, 1998). Egg volume (**ev**) was calculated as a prolate spheroid, $\mathbf{ev} = (1/6)\pi \times \mathbf{ew}^2 \times \mathbf{el}$. Relative egg volume (**rev**) is the ratio of **ev** to $\mathbf{tl}^3$, $\mathbf{rev} = \mathbf{ev}/\mathbf{tl}^3$.

Two related expressions use egg size, egg number and body size to (1) calculate *relative* reproductive effort (**rev** × **ova**) and (2) model the *absolute* reproductive effort as a function of thorax length

$$\mathbf{rev} \times \mathbf{ova} = \frac{\mathbf{ev} \times \mathbf{ova}}{\mathbf{tl}^3} \tag{1}$$

and

$$\mathbf{ev} \times \mathbf{ova} = \alpha \times \mathbf{tl}^{\beta_1} \tag{2}$$

In the first expression the scaling relationship obtained by dividing reproductive volume by thorax length ($\mathbf{tl}^3$ = fly volume) is assumed to result in a non-dimensional factor (volume/volume), i. e., relative reproductive allocation (**rev** × **ova**). If the power = 3 for **tl** assumption is relaxed, then equation (1) can be modified to model constant reproductive effort as a power function of thorax length (**tl**).

The trade-off between egg size and egg number can be modeled as a power function where relative egg volume (**rev**) is a function of ovariole number (**ova**),

$$\mathbf{rev} = \alpha \times \mathbf{ova}^{\beta_0} \tag{3}$$

and

$$\mathbf{ev} = \alpha \times \mathbf{ova}^{\beta_0} \times \mathbf{tl}^{\beta_1} \tag{4}$$

Similar to the conceptual connection of expression (1) and Model 2 the scaling relationship for female body size is assumed to be $\mathbf{tl}^3$ (volume/volume) in Model 3 or can be treated as a fitted constant (β_1) in Model 4. Model 3 and expression (1) are directly related when the trade-off between egg size and ovariole number is isometric. In this case $\beta_0 = -1$ and the constant (α) of Model 3 is equal to **rev** × **ova**.

The models expressed in (2), (3) and (4) can be transformed to linear forms by taking logarithms of both sides of the equations.

$$\log(\mathbf{ev} \times \mathbf{ova}) = \log(\alpha) + \beta_1 \times \log(\mathbf{tl}), \tag{5}$$

$$\log(\mathbf{rev}) = \log(\alpha) + \beta_0 \times \log(\mathbf{ova}), \tag{6}$$

and

$$\log(\mathbf{ev}) = \log(\alpha) + \beta_0 \times \log(\mathbf{ova}) + \beta_1 \times \log(\mathbf{tl}). \tag{7}$$

These linear equations express the relationship between reproductive volume (**ev** × **ova**), relative egg volume (**rev**), and egg volume (**ev**) as respective functions of thorax length (**tl**), ovariole number (**ova**), or both **ova** and **tl**. When the relationships are used in analysis of covariance, regional, intergeneric, interspecific, intraspecific geographic and intraspecific temporal variability in reproductive activity can be investigated.

Comparisons of estimates of **rev** × **ova** and β_0 and β_1 were made for geographic regions and for genera, using data of each species. *Scaptodrosphila hibsici* and *S. aclinata* data were used to evaluate closely related species, intraspecific geographic, temporal, among isofemale line, and laboratory versus field variability. The laboratory versus field comparisons were made by collecting females of *S. hibisci* from one site (Bellingen, N.S.W.) over a two week period in the spring of two successive years (1996 and 1997). Females and new flowers with eggs were collected and returned to the laboratory. The field females were dissected immediately. The flowers were placed on damp sand in jars and incubated at different temperatures. Females emerging from the flowers were allowed to mature in cages with fresh *Hibiscus heterophyllus* blossoms before dissection.

Comparisons for Family, Regions or Genera, and Species used genus means, species means, and population means, respectively, while intraspecific analysis employed site, population or isofemale-line means of *S. hibisci*.

Only species statistics were used for the Hawaiian data because these values were obtained from the literature (Kambysellis and Heed, 1971) and individual data were not available. Because the North American species were mainly from laboratory populations, with low sample sizes, only species means were used for analysis. Throughout the analysis, the classification of genera proposed by Grimaldi (1990) was followed.

ANOVAs (SAS, Proc GLM) employed Type III sums of squares with all nested components considered to be random effects. In all regression analyses least-squares estimation procedures (SAS, Proc REG) were used to estimate β_0 and β_1 and the corresponding r^2 for each model expressed in equations (2), (3) and (4). Reduced major axis estimates of the functional relationship (Rayner, 1985) can be derived from the statistics presented in the tables and results.

Before comparative examination of evolutionary relationships between characters, it is preferable to control for phylogenetic effects (Felsenstein, 1985; Harvey and Pagel, 1991). Because phylogenetic relationships are unknown for the Hawaiian and Australasian taxa, character relationships were examined and compared across several hierarchical levels. Additionally for the North American taxa, Felsenstein's (1985) method of phylogenetically independent contrasts was employed, which provides statistical independence of data points. Independent contrasts were computed (using the phylogenetic topology and branch lengths presented in Figs. 2 and 3 of Pitnick et al., 1999) using the Comparative Analysis by Independent Contrasts (CAIC) program of Purvis and Rambaut (1995). The analyses presented employ a model that assumes gradual evolutionary change in variables, with branch lengths equal to estimated times of divergence (Felsenstein, 1985).

The phylogeny was compiled from a number of sources. The higher level relationships were inferred from several published morphological (Grimaldi et al., 1992; Throckmorton, 1962, 1975) and molecular (Beverley and Wilson, 1982, 1984; Spicer, 1988; Sullivan et al., 1990; Caccone et al., 1992; DeSalle, 1992; Pelandakis and Solignac, 1993; Kwiatowski et al., 1994; Russo et al., 1995; Powell and DeSalle, 1995; Powell, 1997) data sets. In addition to the published sources, an unpublished data set consisting of 2.7 Kb of nuclear large-subunit (28S) ribosomal RNA sequence was used (C. Bell, C. Saux, and G. S. Spicer, unpublished). The lower level relationships were determined both from published sources and from unpublished DNA sequences comprising about 1.5 Kb of the mitochondrial cytochrome oxidase subunits (G. S. Spicer, unpublished). Phylogenetic relationships for the *D. melanogaster* (Ashburner, 1989) and *D. quinaria* (Spicer and Jaenike, 1996) species groups were inferred entirely from the literature, whereas the relationships within the *D. virilis* (Spicer, 1991, 1992) and *D. repleta*

(Wasserman, 1992; Spicer and Pitnick, 1996) species groups were determined by using a combination of published phylogenies and the unpublished sequencing studies. Relationships within the *D. nannoptera* and *D. melanica* species groups were inferred entirely from unpublished sequencing studies.

RESULTS

The species used in the geographic and generic comparisons and associated statistics for thorax length, ovariole number, egg width and egg length are listed in Table 1. The means in Table 1 were used to estimate regional statistics (mean and coefficient of variation, CV) for thorax length, ovariole number and egg volume (Table 2). Comparisons of these regional means show that the Hawaiian region has the largest CVs for all variables, and has the largest mean thorax length and egg volume. The other two geographic regions have similar but smaller mean egg volumes and thorax lengths. The Australasian region has greater variation for ovariole number and egg volume but similar variability for thorax size, when compared to the North American region. However, the means of all three variables for Australasian and North American regions are not statistically different ($\alpha = 0.05$). The only mean that is similar across all three regions is mean ovariole number.

The means in Table 1 also were used to estimate means and CVs for the six genera for which more than one species was examined (Table 2). Analysis of variance shows that all variables (**tl**, **ova** and **ev**) are significantly different among genera. Species belonging to the genus *Idiomyia* are larger, whereas members of the *Scaptomyza* have fewer ovarioles and larger eggs than members of the other genera. The regional differences in body and egg size, setting Hawaii apart from the other regions, is primarily a result of the fact that *Idiomyia* and *Scaptomyza* are restricted to the Hawaiian fauna.

The range in absolute egg size for the species from all regions (Fig. 1) is almost 100× with eggs of *Mycodrosophila variata* from New Caledonia being the smallest (0.0029 mm^3) and eggs of *Scaptomyza undulata* from Hawaii being the largest (0.2691 mm^3). The range in egg size within regions is smaller (18.6× for Hawaiian, 13.3× for Australasian and 3.2× for North American taxa). The range in relative egg size (**REV**) is even more pronounced for the Hawaiian species (93.4×), whereas **REV** for the other two regions is similar to the ranges in absolute egg size (11.1× for Australiasia and 4.3× for North America).

TABLE 1. Geographic Region (R), Genus(G), Species. Mean ± std: Thorax Length (tl, mm), Ovariole Number (ova), Egg Width (ew, mm) and Egg Length (el, mm)

R	G	species	n	tl ± std	ova ± std	n	ew ± std	el ± std
ha	Sz	*caliginosa*	24	0.93 ± 0.078	2.50 ± 0.293	10	0.32 ± 0.029	0.88 ± 0.016
ha	Sz	*mauiensis*	18	0.87 ± 0.034	4.00 ± 0.000	10	0.28 ± 0.015	0.74 ± 0.022
ha	Sz	*oahuensis*	19	0.86 ± 0.042	4.05 ± 0.250	15	0.25 ± 0.016	0.73 ± 0.033
ha	Sz	*throckmortoni*	29	0.84 ± 0.059	4.07 ± 0.253	15	0.28 ± 0.014	0.73 ± 0.034
ha	Sz	*undulata*	6	2.00 ± 0.078	2.00 ± 0.000	15	0.71 ± 0.015	1.02 ± 0.023
ha	Sz	*nasalis*	22	1.90 ± 0.342	4.68 ± 1.660	15	0.48 ± 0.013	1.41 ± 0.038
ha	Sz	*reducta*	4	1.68 ± 0.150	5.33 ± 1.110	0	. .	. .
ha	Sz	*crassifemur*	14	2.06 ± 0.292	5.36 ± 0.902	15	0.39 ± 0.019	1.01 ± 0.091
ha	Sz	*inflatus*	3	1.69 ± 0.025	7.25 ± 0.213	0	. .	. .
ha	Id	*villosus*	5	2.92 ± 0.087	8.40 ± 0.456	30	0.47 ± 0.017	1.66 ± 0.063
ha	Id	*aduncus*	21	2.94 ± 0.119	11.00 ± 1.233	30	0.46 ± 0.022	1.61 ± 0.039
ha	Id	sp	10	2.33 ± 0.129	14.30 ± 0.642	25	0.37 ± 0.012	1.25 ± 0.033
ha	Id	*diamphidiopodus*	18	2.28 ± 0.131	18.44 ± 1.065	25	0.29 ± 0.016	1.09 ± 0.039
ha	Id	*prodita*	12	1.37 ± 0.138	9.00 ± 2.311	0	. .	. .
ha	Id	*trichetosa*	9	1.35 ± 0.096	10.33 ± 1.632	0	. .	. .
ha	Id	*disticha*	244	1.45 ± 0.101	11.79 ± 1.532	30	0.26 ± 0.008	0.90 ± 0.028
ha	Id	*pectinitarsus*	12	1.26 ± 0.028	12.42 ± 1.604	25	0.22 ± 0.007	0.69 ± 0.033
ha	Id	*kambysellisi*	26	1.51 ± 0.076	15.00 ± 1.226	30	0.23 ± 0.007	0.79 ± 0.021
ha	Id	*petalopeza*	16	1.75 ± 0.072	17.75 ± 1.820	0	.	.
ha	Id	*mimica*	325	1.78 ± 0.184	23.85 ± 4.214	39	0.22 ± 0.006	0.74 ± 0.020
ha	Id	*primaeva*	7	3.00 ± 0.124	101.33 ± 8.439	30	0.23 ± 0.006	0.83 ± 0.021
ha	Id	*attigua*	2	2.67 ± 0.063	43.00 ± 1.003	27	0.22 ± 0.009	0.81 ± 0.023
ha	Id	*setosimentum*	33	2.13 ± 0.086	35.61 ± 3.889	24	0.23 ± 0.004	0.86 ± 0.021
ha	Id	*adiastola*	21	2.41 ± 0.092	45.92 ± 7.246	25	0.23 ± 0.006	0.82 ± 0.023
ha	Id	*truncipenna*	6	3.22 ± 0.129	48.00 ± 4.896	25	0.25 ± 0.090	0.96 ± 0.035

ha	Id	*clavisetae*	7	2.71 ± 0.114	38.17 ± 3.558	25	0.31 ± 0.010	0.99 ± 0.031
ha	Id	*picticornis*	9	1.77 ± 0.098	27.44 ± 3.654	33	0.24 ± 0.011	0.81 ± 0.026
ha	Id	*melanocephala*	6	3.31 ± 0.342	36.60 ± 9.243	30	0.22 ± 0.007	0.90 ± 0.034
ha	Id	*silvestris*	13	3.16 ± 0.129	52.38 ± 2.672	36	0.25 ± 0.015	0.94 ± 0.030
ha	Id	*nigribasis*	5	2.98 ± 0.305	53.33 ± 1.886	24	0.24 ± 0.007	1.04 ± 0.024
ha	Id	*pilimana*	8	2.19 ± 0.167	45.00 ± 3.739	25	0.22 ± 0.035	0.75 ± 0.014
ha	Id	*fasciculisetae*	9	2.65 ± 0.222	47.22 ± 6.357	25	0.22 ± 0.007	0.77 ± 0.023
ha	Id	*punalua*	10	2.30 ± 0.136	34.00 ± 3.267	30	0.23 ± 0.011	0.92 ± 0.012
ha	Id	*crucigera*	16	2.09 ± 0.188	40.00 ± 4.288	30	0.21 ± 0.008	0.81 ± 0.021
ha	Id	*engyochracea*	13	2.48 ± 0.115	59.73 ± 7.448	24	0.23 ± 0.014	0.87 ± 0.036
ha	Id	*sproati*	42	2.78 ± 0.149	65.55 ± 5.929	24	0.19 ± 0.008	0.87 ± 0.020
ha	Id	*sejuncta*	7	2.24 ± 0.212	56.83 ± 5.895	21	0.26 ± 0.023	0.97 ± 0.021
ha	Id	*ochracea*	5	2.40 ± 0.096	38.00 ± 4.595	20	0.21 ± 0.007	1.09 ± 0.020
ha	Id	*murphyi*	21	2.43 ± 0.119	41.57 ± 4.798	24	0.21 ± 0.008	0.91 ± 0.013
ha	Id	*villosipedis*	6	2.09 ± 0.144	47.00 ± 3.633	30	0.22 ± 0.006	0.87 ± 0.028
ha	Id	*preapicala*	2	1.20 ± .	37.50 ± .	15	0.20 ± .	0.69 ± .
au	Dr	*ananasse*	6	0.86 ± 0.053	23.00 ± 3.347	2	0.15 ± 0.003	0.48 ± 0.014
au	Dr	*bipectinata*	4	0.92 ± 0.058	25.00 ± 1.414	3	0.15 ± 0.002	0.45 ± 0.020
au	Dr	*ficusphila*	6	0.95 ± 0.086	13.33 ± 3.777	5	0.19 ± 0.013	0.56 ± 0.021
au	Dr	*flavohirta*	4	0.82 ± 0.075	22.00 ± 3.559	1	0.17 ± .	0.53 ± .
au	Dr	*pseudotetrachaeta*	6	1.26 ± 0.068	27.50 ± 2.168	5	0.20 ± 0.008	0.57 ± 0.046
au	Dr	*sulfurigaster*	6	1.24 ± 0.096	37.67 ± 5.715	2	0.18 ± 0.000	0.58 ± 0.003
au	Hi	*hannae*	6	1.31 ± 0.088	36.00 ± 4.427	2	0.20 ± 0.003	0.53 ± 0.008
au	Hi	*hirudo*	3	1.10 ± 0.047	24.00 ± 5.568	1	0.18 ± .	0.56 ± .
au	Hi	*laurelae*	1	1.13 ± .	24.00 ± .	1	0.17 ± .	0.40 ± .
au	Hi	*mixtura*.1	6	1.11 ± 0.125	34.17 ± 7.910	1	0.18 ± .	0.43 ± .
au	Hi	*polypori*	3	1.07 ± 0.114	24.33 ± 2.887	0	. .	. .
au	Hi	*trifurca*	2	1.14 ± 0.042	6.50 ± 0.707	1	0.30 ± .	0.80 ± .
au	My	*boudinoti*	6	0.93 ± 0.094	22.00 ± 4.427	3	0.19 ± 0.010	0.46 ± 0.033
au	My	*cf.boudinoti*	7	0.73 ± 0.043	24.14 ± 4.706	2	0.12 ± 0.000	0.41 ± 0.000

Continued

TABLE 1. *Continued.*

R	G	species	n	tl ± std	ova ± std	n	ew ± std	el ± std
au	My	*cf.minor*	8	1.06 ± 0.060	28.25 ± 1.909	3	0.15 ± 0.013	0.45 ± 0.015
au	Mi	*elator*	4	0.81 ± 0.039	22.25 ± 2.500	2	0.14 ± 0.011	0.43 ± 0.017
au	My	*minor*	4	1.11 ± 0.055	27.25 ± 5.620	6	0.17 ± 0.005	0.48 ± 0.016
au	Zy	*samoensis*	6	1.04 ± 0.117	27.50 ± 4.278	2	0.18 ± 0.011	0.54 ± 0.022
au	Sc	*bryani*	3	0.85 ± 0.051	71.33 ± 9.238	3	0.13 ± 0.005	0.35 ± 0.018
au	Sc	*collessi*	5	1.10 ± 0.062	29.80 ± 3.962	1	0.17 ± .	0.55 ± .
au	Sc	*eluta*	8	0.98 ± 0.062	32.75 ± 5.092	2	0.17 ± 0.011	0.44 ± 0.006
au	Sc	*brunnea*	3	1.46 ± 0.047	64.33 ± 5.686	5	0.18 ± 0.010	0.49 ± 0.027
au	Sc	*fumida*	4	1.06 ± 0.064	45.25 ± 5.679	2	0.16 ± 0.006	0.46 ± 0.008
au	Sc	*fuscithorax*	1	0.98 ± .	37.00 ± .	0	. .	. .
au	Sc	*inornata*	3	1.04 ± 0.088	47.67 ± 13.80	3	0.18 ± 0.008	0.53 ± 0.016
au	Sc	*lattivittata*	6	1.20 ± 0.077	36.50 ± 4.930	3	0.19 ± 0.005	0.53 ± 0.008
au	Sc	*novoguinensis*	5	1.25 ± 0.127	74.40 ± 9.529	2	0.14 ± 0.000	0.44 ± 0.017
au	Sc	*rhabdote*	1	1.14 ± .	38.00 ± .	0	. .	. .
au	Sc	*zophera*	8	1.01 ± 0.083	47.13 ± 8.806	2	0.14 ± 0.011	0.45 ± 0.028
au	Sc	Scapto.b	6	0.99 ± 0.111	45.83 ± 8.134	1	0.15 ± .	0.43 ± .
au	Sc	*hibisci*	787	0.91 ± 0.098	14.60 ± 4.548	443	0.21 ± 0.020	0.66 ± 0.050
au	Sc	*aclinata*	286	0.77 ± 0.083	13.85 ± 3.757	181	0.20 ± 0.015	0.63 ± 0.037
na	Dr	*floricola*	46	0.88 ± 0.073	14.07 ± 1.855	30	0.19 ± 0.012	0.59 ± 0.029
na	Dr	*repleta*	8	1.32 ± 0.033	36.50 ± 3.295	8	0.16 ± 0.005	0.52 ± 0.022
na	Dr	*arizonae*	10	1.05 ± 0.025	34.60 ± 2.413	6	0.15 ± 0.008	0.46 ± 0.019
na	Dr	*mettleri*	8	1.20 ± 0.030	44.75 ± 1.832	7	0.14 ± 0.003	0.49 ± 0.010
na	Dr	*micromettleri*	7	1.14 ± 0.010	36.29 ± 2.563	4	0.17 ± 0.007	0.46 ± 0.017
na	Dr	*mojavensis*	6	1.03 ± 0.029	33.17 ± 2.639	7	0.15 ± 0.005	0.46 ± 0.011
na	Dr	*acanthoptera*	8	1.15 ± 0.028	41.88 ± 2.949	8	0.16 ± 0.004	0.53 ± 0.014
na	Dr	*nannoptera*	10	1.06 ± 0.026	37.60 ± 1.647	6	0.17 ± 0.009	0.48 ± 0.018

na	Dr	*wassermani*	8	1.15 ± 0.028	33.88 ± 2.949	8	0.16 ± 0.002	0.50 ± 0.011
na	Dr	*pachea*	6	0.99 ± 0.045	28.17 ± 2.317	5	0.16 ± 0.007	0.45 ± 0.022
na	Dr	*americana*	10	1.29 ± 0.024	31.60 ± 2.716	10	0.17 ± 0.005	0.54 ± 0.013
na	Dr	*borealis*	9	1.29 ± 0.036	30.78 ± 1.563	9	0.18 ± 0.008	0.53 ± 0.023
na	Dr	*lummei*	8	1.42 ± 0.035	36.00 ± 1.927	7	0.18 ± 0.005	0.57 ± 0.009
na	Dr	*virilis*	14	1.25 ± 0.041	41.21 ± 2.914	14	0.17 ± 0.006	0.56 ± 0.019
na	Dr	*lacicola*	6	1.22 ± 0.029	28.67 ± 1.211	4	0.16 ± 0.007	0.49 ± 0.006
na	Dr	*montana*	10	1.32 ± 0.027	28.80 ± 1.476	8	0.18 ± 0.004	0.58 ± 0.015
na	Dr	*texana*	5	1.27 ± 0.032	38.00 ± 2.739	5	0.16 ± 0.002	0.51 ± 0.010
na	Dr	*eohydei*	9	1.28 ± 0.041	39.22 ± 3.667	9	0.16 ± 0.007	0.53 ± 0.021
na	Dr	*nigrohydei*	6	1.34 ± 0.041	47.33 ± 5.203	6	0.17 ± 0.005	0.52 ± 0.026
na	Dr	*hydei*	8	1.43 ± 0.054	51.75 ± 4.268	8	0.17 ± 0.006	0.53 ± 0.019
na	Dr	*bifurca*	15	1.53 ± 0.052	51.53 ± 4.454	13	0.17 ± 0.009	0.54 ± 0.015
na	Dr	*robusta*	4	1.47 ± 0.048	41.25 ± 4.573	3	0.16 ± 0.002	0.53 ± 0.003
na	Dr	*immigrans*	8	1.48 ± 0.044	65.38 ± 5.041	8	0.16 ± 0.007	0.53 ± 0.012
na	Dr	*micromelanica*	7	1.16 ± 0.018	25.57 ± 1.902	5	0.18 ± 0.006	0.51 ± 0.016
na	Dr	*melanica*	5	1.33 ± 0.014	42.20 ± 1.095	3	0.20 ± 0.005	0.50 ± 0.019
na	Dr	*neotestacae*	5	1.06 ± 0.069	36.60 ± 3.647	5	0.16 ± 0.006	0.46 ± 0.014
na	Dr	*putrida*	2	1.04 ± 0.011	40.00 ± 1.414	2	0.15 ± 0.006	0.45 ± 0.025
na	Dr	*subpalustris*	5	1.35 ± 0.016	26.20 ± 1.095	5	0.20 ± 0.005	0.54 ± 0.011
na	Dr	*melanogaster*	7	0.98 ± 0.036	33.14 ± 1.345	7	0.18 ± 0.007	0.49 ± 0.022
na	Dr	*simulans*	6	0.89 ± 0.024	36.83 ± 2.563	7	0.17 ± 0.003	0.49 ± 0.013
na	Dr	*willistoni*	5	0.90 ± 0.030	22.60 ± 1.673	5	0.17 ± 0.008	0.48 ± 0.022
na	Dr	*pseudoobscura*	12	1.09 ± 0.057	45.42 ± 5.915	7	0.16 ± 0.006	0.46 ± 0.016
na	Dr	*busckii*	7	0.98 ± 0.081	52.86 ± 9.771	4	0.13 ± 0.008	0.38 ± 0.012
na	Hi	*duncani*	6	1.17 ± 0.037	47.00 ± 8.075	5	0.16 ± 0.006	0.48 ± 0.016
na	My	*claytonae*	5	1.08 ± 0.056	31.20 ± 4.382	4	0.17 ± 0.009	0.48 ± 0.016
na	My	*dimidiata*	1	0.97 ± .	28.00 ± .	1	0.17 ± .	0.50 ± .

R: ha = Hawaiian Islands, au = Australia & New Caledonia, na = North America.
G: Dr = *Drosophila*, Hi = *Hirtodrosophila*, Id = *Idiomyia*, Mi = *Microdrosophila*, My = *Mycodrosophila*, Sc = *Scaptodrosophila*, Sz = *Scaptomyza*, Zy = *Zygothrica*.

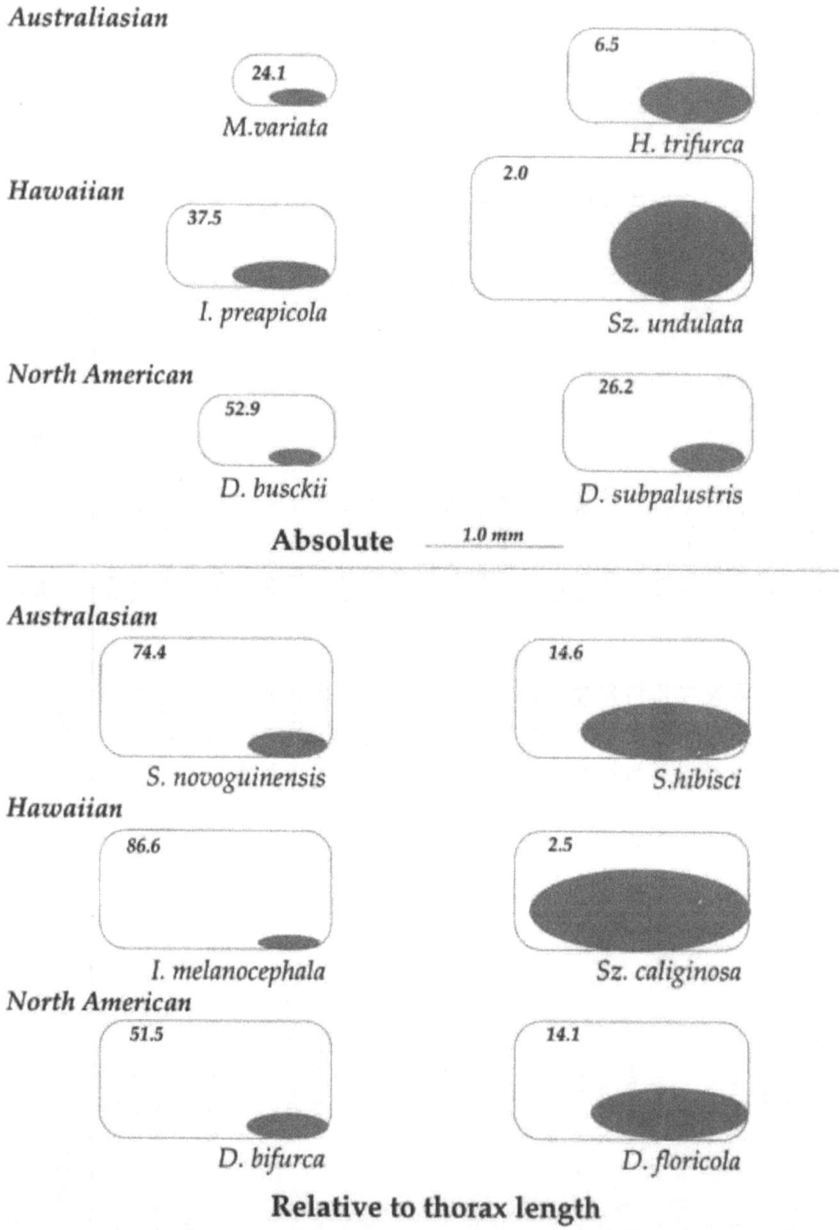

FIG. 1. Range (low, left to high, right) in egg size for species in the three geographic regions. All figures show egg size as shaded ovals with width < length. The top three comparisons have eggs in open boxes drawn with thorax length as the base and 1/3 thorax length as the height, on an absolute scale. The bottom three comparisons use a relative scale. Mean ovariole number for each species is displayed in the upper left for each case.

TABLE 2. Mean and Coefficient of Variation (CV) for Thorax Length (tl, mm), Ovariole Number and Egg Volume (ev, mm^3) of Species in the Three Geographic Regions

Region	n	tl	(CV)	ova	(CV)	ev	(CV)
Australia	29	1.04^b	(17.20)	32.69^a	(50.46)	0.0089^b	(73.48)
Hawaii	36	2.17^a	(32.33)	32.95^a	(74.07)	0.0492^a	(119.45)
North American	36	1.18^b	(15.05)	37.22^a	(26.28)	0.0074^b	(22.01)
ANOVA Region ($F_{2,98}$)		66.3***		0.69		15.75***	
Genus							
Drosophila	39	1.16^a	(16.55)	35.44^a	(29.86)	0.0076^a	(24.85)
Hirtodrosophila	6	1.16^a	(6.67)	$28.61^{a,b}$	(48.30)	0.0135^a	(93.54)
Idiomyia	29	2.37^b	(24.30)	39.98^a	(54.84)	0.0386^a	(111.79)
Microdrosophila	1	0.81^a		$22.25^{a,b}$		0.0044^a	
Mycodrosophila	6	0.98^a	(14.36)	$26.81^{a,b}$	(12.17)	0.0063^a	(30.64)
Scaptodrosophila	12	1.05^a	(18.86)	43.62^a	(44.96)	0.0079^a	(46.84)
Scaptomyza	7	1.35^a	(44.15)	3.81^b	(30.90)	0.0930^b	(100.01)
Zygothrica	1	1.04^a		$27.50^{a,b}$		0.0091^a	
ANOVA Genus ($F_{7,93}$)		31.67***		5.35***		7.09***	

***$P < 0.001$.
[a,b] represent statistically similar groups for each variable.

The range in ovariole number reflects the egg size variation for each region. The Hawaiian flies range from 2 to 101 ovarioles. The Australasian species had a low of 6.5 and a high of 71.3 ovarioles. These two ranges are much larger than for the North American species (range: 14.1 to 65.4).

Reproductive Effort

Table 3 reports the analysis of reproductive effort considering the following two assumptions: 1) *The scaling relationship with thorax length is constant* (tl^3). In this case change in volume of reproductive tissue is relative to the volume of the female. Values of **rev** × **ova** of Table 3 differ at all levels of the analysis; regions are different, genera are different, species are different, and populations are different. 2) *The scaling relationship with thorax length is constant but not necessarily* tl^3. In this case the scaling relationship is estimated for each comparison. Values of **ev** × **ova** of Table 3 are the same in each region and for different populations of *S. hibisci* but differ for genera and the closely related *Scaptodrosophila* species. Comparison of **rev** × **ova** and **ev** × **ova** shows that the rank order of these measures of reproductive effort is the same for regions and species, but with a major shift in

TABLE 3. Analysis of Reproductive Effort for Regions, Genera, 2 Species within Australia and Sites for *S. hibisci*: **rev.ova, ev.ova, rev** and **ev** Were Evaluated Using Equation 1 and Models 2, 3 and 4. Anti-logarithms of Least-squares Means from the ANOVAs Are Given. If Slopes Were Homogenous the Covariate*Effect Interaction Was Not Included in the Model

		Model			
		1	2	3	4
Dependent		**rev.ova**	**ev.ova**	**rev**	**ev**
Covariate			**tl**	**ova**	**ova, tl**
REGION	(species)				
Australian	29	0.219	0.398	0.0140	0.0129
Hawaiian	36	0.098	0.334	0.0058	0.0137
North American	36	0.169	0.356	0.0112	0.0116
ANOVA					
Region F		29.78***	2.12	51.96***	2.72
(dfn, dfd)		(2, 98)	(2, 97)	(2, 97)	(2, 96)
Homogenity of Slopes (P)			0.12	0.22	0.26, 0.08
GENUS	(species)				
Drosophila	39	0.172	0.344	0.0113	0.0117
Hirtodrosophila	6	0.169	0.349	0.0110	0.0124
Idiomyia	29	0.090	0.384	0.0056	0.0140
Microdrosophila	1	0.188	0.256	0.0123	0.0089
Mycodrosophila	6	0.179	0.300	0.0118	0.0104
Scaptodrosophila	12	0.268	0.467	0.0178	0.0155
Scaptomyza	7	0.132	0.272	0.0063	0.0120
Zygothrica	1	0.225	0.417	0.0151	0.0144
ANOVA					
Genus F		11.94***	4.74***	17.22***	3.14**
(dfn, dfd)		(7, 95)	(7, 93)	(7, 93)	(7, 92)
Homogenity of Slopes (P)			0.32	0.92	0.58, 0.09
SPECIES	(sites)				
S. aclinata	15	0.382	0.211	0.0554	0.0120
S. hibisci	34	0.256	0.174	0.0367	0.0150
ANOVA					
Species F		49.72***	1.85**	48.91***	4.19*
(dfn, dfd)		(1, 47)	(1, 46)	(1, 46)	(1, 45)
Homogenity of Slopes (P)			0.03*	0.51	0.63, 0.82
SITES (*S. hibisci* isolines)					
BCK	9	0.318	0.256	0.0163	0.0131
BEL	9	0.371	0.274	0.0197	0.0134
TRD	5	0.350	0.276	0.0177	0.0136
ANOVA					
Sites F		4.71*	1.02	5.06*	0.63
(dfn, dfd)		(1, 20)	(2, 19)	(2, 19)	(1, 18)
Homogenity of Slopes (P)			0.22	0.89	0.27, 0.29

*P < 0.05, **P < 0.01, ***P < 0.001.

TABLE 4. Analysis of Reproductive Effort at One Site (Bellingen, N.S.W.) over Two Weeks in Each of Two Years: **rev.ova, ev.ova, rev** and **ev** Were Evaluated Using Equation 1 and Models 2, 3 and 4. Anti-Logarithms of Least-Squares Means from the ANOVAs Are Given

		Model			
		1	2	3	4
Dependent		**rev.ova**	**ev.ova**	**rev**	**ev**
Covariate			**tl**	**ova**	**ova, tl**
Year: Week/Strain (females)					
1996:					
Oct. 10/Field	14	0.389	0.287	0.0444	0.0200
Lab.18	2	0.281	0.207	0.0322	0.0111
Lab.21.5	4	0.340	0.230	0.0392	0.0132
Lab.25	3	0.316	0.225	0.0357	0.0132
Oct. 17/Field	36	0.375	0.281	0.0424	0.0181
Lab.18	8	0.323	0.247	0.0361	0.0138
Lab.21.5	11	0.283	0.221	0.0320	0.0130
Lab.25	8	0.301	0.232	0.0342	0.0134
1997:					
Oct. 22/Field	13	0.433	0.299	0.0495	0.0176
Lab.25	3	0.381	0.254	0.0445	0.0167
Lab.29	16	0.391	0.239	0.0445	0.0144
Nov. 1/Field	30	0.573	0.383	0.0654	0.0226
Lab.18	17	0.358	0.262	0.0403	0.0155
Lab.25	12	0.349	0.239	0.0393	0.0140
Lab.29	23	0.343	0.224	0.0382	0.0135
ANOVA					
Year	F	206.7***	20.23*	56.80*	65.72*
	(dfn, dfd)	(1, 2)	(1, 2)	(1, 2)	(1, 2)
Week/Year	F	0.01	0.02	0.03	0.01
	(dfn, dfd)	(2, 11)	(2, 11)	(2, 11)	(2, 11)
Strain/Week/Year	F	12.90***	12.54***	11.30***	25.5***
	(dfn, dfd)	(11, 185)	(11, 184)	(11, 184)	(11, 183)

$*P < 0.05$, $**P < 0.01$, $***P < 0.001$.

the generic order for *Idiomyia* which moves from the lowest effort to the third highest. A similar analysis for field versus laboratory females collected at the same site during two weeks of two years (Table 4) shows the reproductive effort was different in the two years, similar within years and higher in field collected females as compared to laboratory reared females.

Variance components analysis of the isofemale line data showed no heritability in **rev** × **ova**. Site variance (0.00090) was eleven times larger than line within site variance (0.00008), while individuals within lines (error) showed the largest contribution (0.00832) to the variance in **rev** × **ova**.

Egg Size

Egg volume was compared by scaling egg volume with constant thorax length cubed and correcting for the trade-off with ovariole number (Model 3). This analysis shows that regional, generic, species and population categories all differ. On relaxing the assumption that the scaling with **tl** is $\mathbf{tl}^3$ (Model 4), the analysis indicates that the regional relative egg volumes are similar and populations of *S. hibisci* are also similar. However, generic and sister species egg volumes are still significantly different (Table 3).

The corresponding egg volume analysis for field versus laboratory females of *S. hibisci* (Table 4) shows that volumes differed from year to year and that laboratory females had smaller eggs than the corresponding field collected females.

Allometry of Egg Size and Egg Number

Models 3 and 4 relate egg size to ovariole number with constant female volume set at $\mathbf{tl}^3$ in Model 3 or estimated as a model component in Model 4. Two comparisons can be made for β_0. The first is whether the trade-off is equivalent for the groups under comparison (Ho: $\beta_{0a} = \beta_{0b} = \beta_{0c} \ldots\ldots$) and the second is whether the trade-off is isometric (Ho: $\beta_0 = -1$). The former comparison is a test of homogeneity of slopes in the Analysis of Covariance (Table 3). In all cases the slopes were homogenous, indicating that the trade-off within regional, family, generic, specific and intraspecific levels was similar for each comparison. Regression analysis for all species indicates the trade-off is isometric ($\hat{\beta}_0 = -0.996 \pm 0.069$, n = 101, $r^2 = 0.675$) when $\mathbf{tl}^3$ was used to scale body size. Similar analysis estimating both β_0 and β_1 (Model 4) show the trade-off to be slightly higher than –1 ($\hat{\beta}_0 = -0.869 \pm 0.034$, $\hat{\beta}_1 = 1.847 \pm 0.062$, n = 101, $r^2 = 0.928$). The estimates for β_0 and β_1 are given for each category in Table 5.

Analysis of species (*S. hibisci* and *S. aclinata*) over all sites showed that the two species have poor fits to either Model 3 or 4 ($r^2 = 0.203$ and 0.220, respectively) with β_0 estimates between 0 and –1 ($\hat{\beta}_0 = -0.484 \pm 0.203$ and $\hat{\beta}_0 = -0.207 \pm 0.093$, respectively). Analysis of populations (isofemale line

TABLE 5. Estimates of β_0 (se) β_1 (se) and r^2 for Geographic, Generic and Species Categories. The Three Rows for each Category Represents Models (3), (2) and (4), Respectively

REGION	n	β_0	(se)	β_1	(se)	r^2
Australian	29	–0.895	(0.118)			0.681
				2.067	(0.316)	0.614
		–0.789	(0.103)	1.881	(0.312)	0.743
Hawaiian	36	–1.152	(0.009)			0.894
				2.092	(0.127)	0.889
		–2.012	(0.144)	1.860	(0.163)	0.870
North American	36	–1.107	(0.180)			0.526
				1.454	(0.202)	0.604
		–0.603	(0.107)	1.098	(0.197)	0.560
Phylogenetic NA	32	–0.724	(0.172)			0.363
				1.824	(0.313)	0.522
		–0.506	(0.131)	1.506	(0.276)	0.539
GENERA (Family)	7	–0.804	(0.159)			0.836
				1.919	(0.218)	0.939
		–0.870	(0.110)	2.208	(0.302)	0.967
Drosophila	39	–1.175	(0.148)			0.631
				1.626	(0.183)	0.681
		–0.645	(0.102)	1.229	(0.197)	0.583
Hirtodrosophila	6	–0.926	(0.172)			0.878
				4.730	(1.717)	0.655
		–0.966	(0.188)	4.630	(2.050)	0.901
Idiomyia	29	–1.131	(0.122)			0.760
				1.879	(0.201)	0.764
		–0.905	(0.095)	1.780	(0.224)	0.807
Mycodrosophila	6	–1.913	(0.832)			0.569
				2.335	(0.682)	0.745
		–1.645	(1.227)	2.665	(0.982)	0.718
Scaptodrosophila	12	–1.140	(0.180)			0.709
				1.460	(0.292)	0.714
		–0.897	(0.110)	1.328	(0.325)	0.882
Scaptomyza	7	–0.941	(0.674)			0.280
				1.925	(0.247)	0.924
		–0.927	(0.343)	1.925	(0.275)	0.933
SPECIES (Australia n = sites)						
S. aclinata	18	–0.646	(0.271)			0.262
				5.355	(1.009)	0.690
		0.089	(1.759)	0.902	(1.759)	0.117
S. hibisci	34	–0.465	(0.168)			0.194
				1.000	(0.871)	0.040
		–0.124	(0.099)	–0.940	(0.282)	0.236
SITES (*S. hibisci* n = isolines)						
BCK	9	–0.904	(0.240)			0.669
				2.143	(0.833)	0.486
		0.402	(0.169)	–1.238	(0.488)	0.521
BEL	9	–1.227	(0.583)			0.388
				3.600	(1.374)	0.530
		0.009	(0.438)	1.042	(0.438)	0.151
TRD	5	–0.893	(0.745)			0.324
				0.436	(1.028)	0.058
		–0.263	(0.328)	–0.757	(0.917)	0.661

*P < 0.05, **P < 0.01, ***P < 0.001.

means of *S. hibisci*) supported an isometric relationship for Model 3 ($\hat{\beta}_0$ = −1.01 ± 0.271, r^2 = 0.40) but did not have significant effects for **tl** or **ova** in Model 4.

Quantitative corrections for phylogenetic effects were only possible for the North American *Drosophila*. The parameter estimates for Model 2 (β_1), Model 3 (β_0) and Model 4 (β_0, β_1) are given in Table 5. Comparison with the comparable results for the North American region without correction shows no significant differences.

DISCUSSION

The analysis of reproductive effort and the trade-off between egg size and egg number is premised on the following ecological, physiological and evolutionary assumptions:

1. **Egg volume is independent of female age, diet, temperature and humidity.** This assumption has not been investigated to any great extent in Drosophilidae (but see Avelar, 1993). Bernardo (1996) discusses the general expectation that propagule size should be affected by maternal condition and her ecological situation. This expectation is supported by analysis comparing the size of eggs from field caught *S. hibisci* females with those of females from the same site following one generation of laboratory rearing from larvae to adult stage (Table 4). Field-caught females had larger eggs (23–60% larger) than laboratory-reared flies, indicating that age (relatively young in the laboratory), climatic differences, or nutrition (unknown in the field) could be important to egg size.
2. **Oviposition opportunities are unlimited.** This assumption also is likely false. Both seasonal and species differences in reliability of oviposition substrate may compromise nearly all levels of comparative analyses. For instance, a prolonged break in oviposition opportunity can cause ovariole activity to cease as mature eggs accumulate in the abdomen (King and Sang, 1959). With respect to species differences, rare oviposition opportunity coupled with unconstrained larval resources when the opportunity is present is believed to select for a reproductive strategy featuring many ovarioles producing small eggs (Montague et al., 1981, Kambysellis et al., 1995). Without knowledge of how body size covaries with substrate availability, it is not possible to know how violation of this assumption influences the egg size/egg number trade-off.

3. **Egg number is equivalent to ovariole number.** We expect this assumption often to be false. Some ovarioles may not be functional and the proportion of nonfunctional ovarioles may increase with female age (Gasser et al., 2000). Our experience with *S. hibsci* is that females rarely have egg numbers equivalent to ovariole numbers in either field or laboratory reared flies. In many cases, only one or very few ovarioles are active. This situation appears to be true also for Hawaiian flower-breeders (Montague, 1984, 1989) and leaf-breeders (Kambysellis et al., 1995). On the other hand, Hawaiian bark-breeders typically have many ovarioles with multiple eggs developing in each ovariole (Kambysellis et al., 1995). Consequently, ovariole number is better viewed as an index of potential instantaneous egg production for each species. Consistent with this viewpoint, David (1970) and Boulétreau-Merle et al. (1982) observed a positive correlation between ovariole number and the maximum daily rate of oviposition (but see Wayne et al., 1997).

   It is also worth noting here that detailed studies of the function of male accessory gland proteins following insemination in *D. melanogaster* have determined that seminal fluid (i.e., the "sex peptide" 26Aa) can enhance the egg production rate of females (Chen et al., 1988; Kalb et al., 1993; Herndon and Wolfner, 1995). This effect is believed to have arisen through sexual conflict over sperm use (Eberhard, 1996; Holland and Rice, 1998). Variation among species or populations in the intensity of postcopulatory sexual selection and in the outcome of sexually antagonistic coevolution (Rice, 1998) could therefore contribute to substantive differences in intrinsic and realized egg production rates.
4. **Instantaneous reproductive effort is a good index of total reproductive effort.** Even if oviposition opportunities are unlimited, females of different species, populations, or samples may not produce per unit volume of egg at the same rate or cost. We examine the egg size-egg number relationship using a static measure of reproduction (i.e., egg size × ovariole number). At best, this variable reliably indicates the instantaneous potential for reproduction. However, an extension to lifetime reproductive effort would require knowledge about the dynamics of egg production and oviposition in relation to oviposition opportunity and longevity of the female. Considering such extensions, the reproductive effort and the parameter α may not be constant for females of different sizes, different senescence patterns (Carlson et al., 1998) or in different ecological situations (Kambysellis et al., 1979). Moreover, true reproductive effort, which is the proportion of energy flowing through

TABLE 6. Goodness of fit Model 2 at Each Level of the Taxonomic Organization

	n	r^2	$\beta_1 \pm$ se
Genera	7	0.94	1.92 ± 0.22
Scaptodrosophila	12	0.71	1.46 ± 0.29
S. hibisci	34	0.04	1.00 ± 0.87
Populations (*S. hibisci*)	23	0.54	1.64 ± 0.30

the organism that is devoted to reproduction, cannot be determined confidently by any static measure (Hirshfield and Tinkle, 1975).

5. **There is a consistent relationship between body size and reproductive effort.** If species differ in their reproductive effort, or in the proportion of their reproductive effort devoted to egg production, then the trade-off between egg size and number may not be discernable. For example, an increase in egg size with no reduction in egg number may not indicate a lack of trade-off between these characters if it is also associated with an increase in reproductive effort. Unfortunately, no criterion is known for quantifying investment in egg production that is independent of egg size and number.
6. **Allocation to reproduction is independent of the trade-off between egg size and egg number.** Most theoretical treatments of egg size evolution have assumed that egg size and total resources devoted to reproduction are optimized independently (Vance, 1973a, 1973b; Smith and Fretwell, 1974; Brockelman, 1975; Lloyd, 1987; McGinley et al., 1987; Sargent et al., 1987; Sinervo et al., 1992; but see Winkler and Wallin, 1987). Once total reproductive allocation is optimized, resources are divided among the maximum number of optimally sized offspring, resulting in a trade-off between offspring size and number. This assumption has been challenged empirically by a recent study by Schwarzkopf et al. (1999) in which selection for increased and decreased egg size was imposed on *D. melanogaster*. Total reproductive allocation did not change in lines selected for large eggs but was reduced in lines selected for small eggs.
7. **Body weight is proportional to the cube of a linear dimension, i.e., thorax length cubed.** This assumption has not been tested for many *Drosophila* species because most studies report the correlation between body weight and thorax length and not with thorax length cubed. We reanalyzed data for *D. hydei* reported by Pitnick and

Markow (1994) and found the regression of log (dry weight) on log (thorax length) had a coefficient less than the expected value of 3 ($\hat{\beta} = 2.47 \pm 0.127$, $n = 20$, $r^2 = 0.954$). However, Robertson and Reeve (1953) argued that the relationship of weight and thorax length cannot be expected to be exact because weight is more variable than thorax length. Their paper reports that females can increase weight by 40% and males 5% during adult life. As a consequence, thorax length is considered to be a more accurate measure of size. Robertson and Reeve (1953) also discuss the functional relationship (Kermack and Haldane, 1951) and the expectation that the regression of log (weight) on log (thorax length) should have a slope of 3. However, Robertson and Reeve estimate the slope to be 1.64 in experiments with males of *D. melanogaster*.

Despite these considerations, our use of thorax length cubed to correct for size differences in expression (1) and Model 3 is a geometric scale correction (volume/volume) and does not take into account differences in specific gravity or the state of tissues (i.e., a mass consideration).

8. **Field caught and laboratory reared females represent the same sample.** Here we compare the egg size/number trade-off between different continental radiations. However, North American flies examined were primarily obtained from laboratory cultures whereas the Hawaiian and Australasian flies were mostly collected from the field. Intraspecific comparison of field-caught and laboratory reared females of *S. hibsici* (Table 4), however, shows the trade-off is essentially the same for both samples even though reproductive effort is generally higher for field-caught individuals (a similar reduction in reproductive effort in the laboratory versus the field has been described for a grasshopper; Kriegsbaum, 1988 as cited in Stearns, 1992). The primary reason for this difference is that egg volume is always larger in field-caught flies than in laboratory reared ones ($F_{3,186} = 75.6$, $P < 0.001$).
9. **Taxa represent independent data for each comparison.** It is recognized that related taxa do not constitute independent data points for statistical purposes (Felsenstein, 1985; Harvey and Pagel, 1991). Nevertheless, because detailed phylogenetic relationships are unknown for the Hawaiian and Australasian taxa, only data for the North American taxa could be properly analyzed using phylogenetically independent contrasts (Harvey and Pagel, 1991). This analysis showed that phylogenetic corrections resulted in similar parameter estimates (Table 5). Thus, for the North American *Drosophila* species the relationships between egg size, ovariole

number and body size have not been subject to significant phylogenetic constraint. Furthermore, the lack of statistical difference between slopes generated by interspecific regression and those generated by intergeneric regression (Figs. 2 and 4) suggest that these data are robust to violations of this assumption.

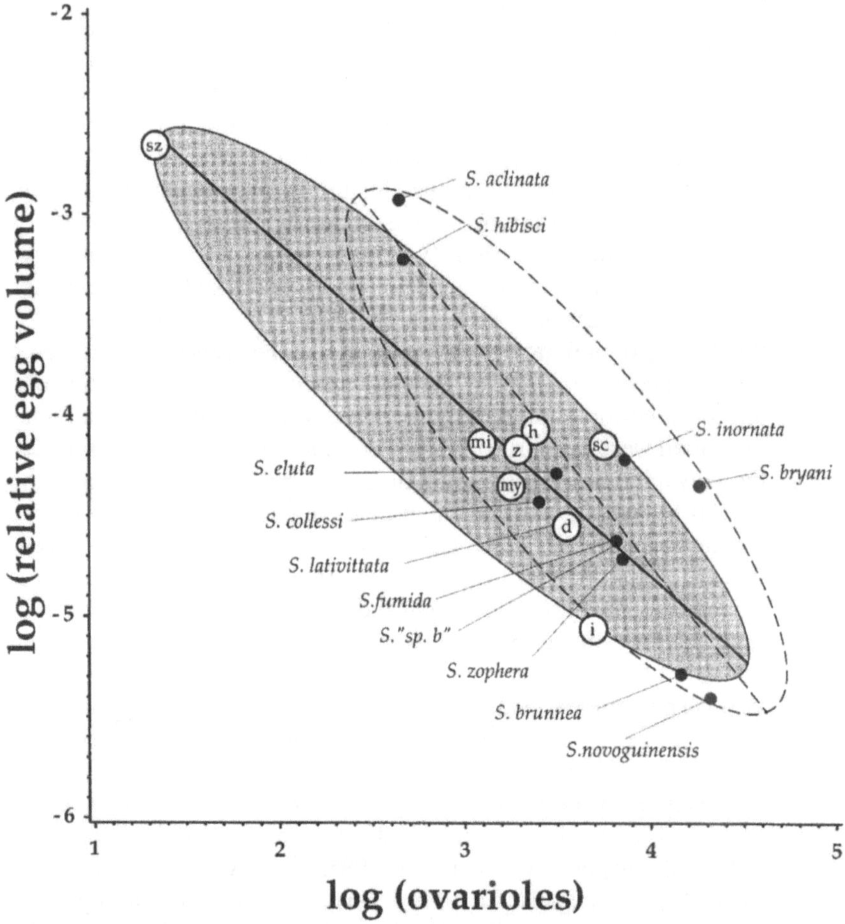

FIG. 2. The trade-off between relative egg size and ovariole number for genera of the Family (circles with generic abbreviations) and for species of *Scaptodrosophila* (solid circles). The range of the family regression line and the variation among genera is shown as a solid line and shaded oval. The genus line and variation is shown as a dashed lined and open dashed oval. sz = *Scaptomyza*, h = *Hirtodrosophila*, sc = *Scaptodrosophila*, z = *Zygothrica*, mi = *Microdrosophila*, my = *Mycodrosophila*, d = *Drosophila*, and i = *Idiomyia*.

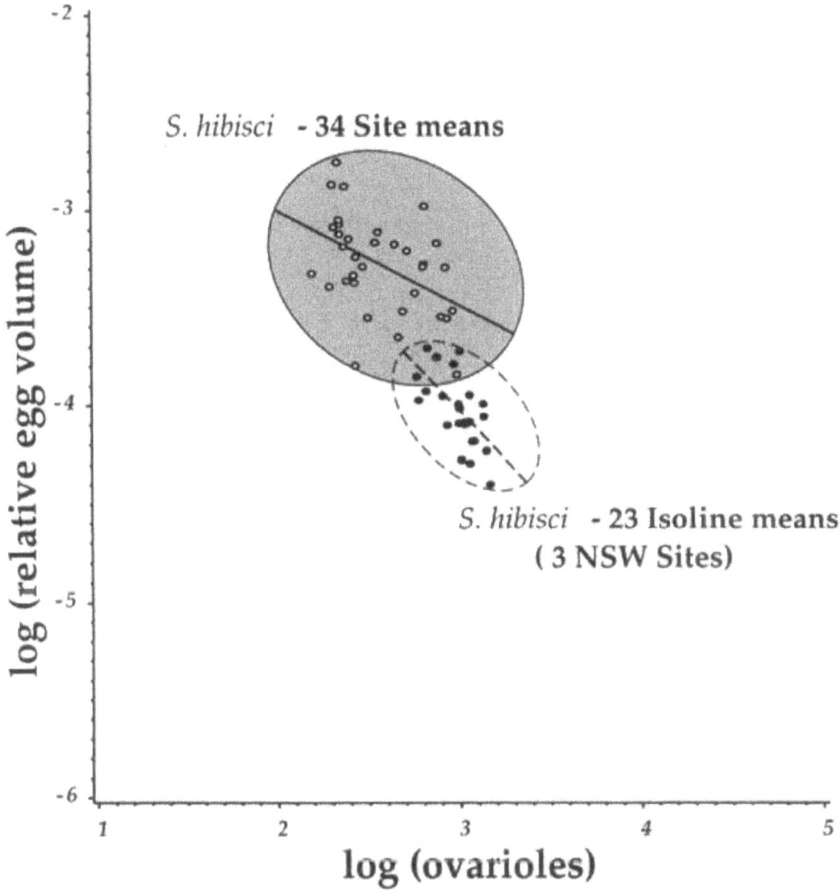

FIG. 3. The trade-off between relative egg size and ovariole number for population means (sites) of *S. hibisci* in eastern Australia (open circles with shaded oval boundary) and for isoline means of three sites in N. S. W. (closed circles with dashed line and dashed oval boundary).

General

The very broad conclusion of the interspecific comparisons is that the relationship between egg size and ovariole number is close to the isometric value, $\beta_0 = -1$, across geographic regions and taxonomic categories (Table 5). However, the $\beta_0 = -1$ slope was not always mirrored in the spatial and temporal analysis of *S. hibisci*, where $\beta_0 > -1$ for the species comparison.

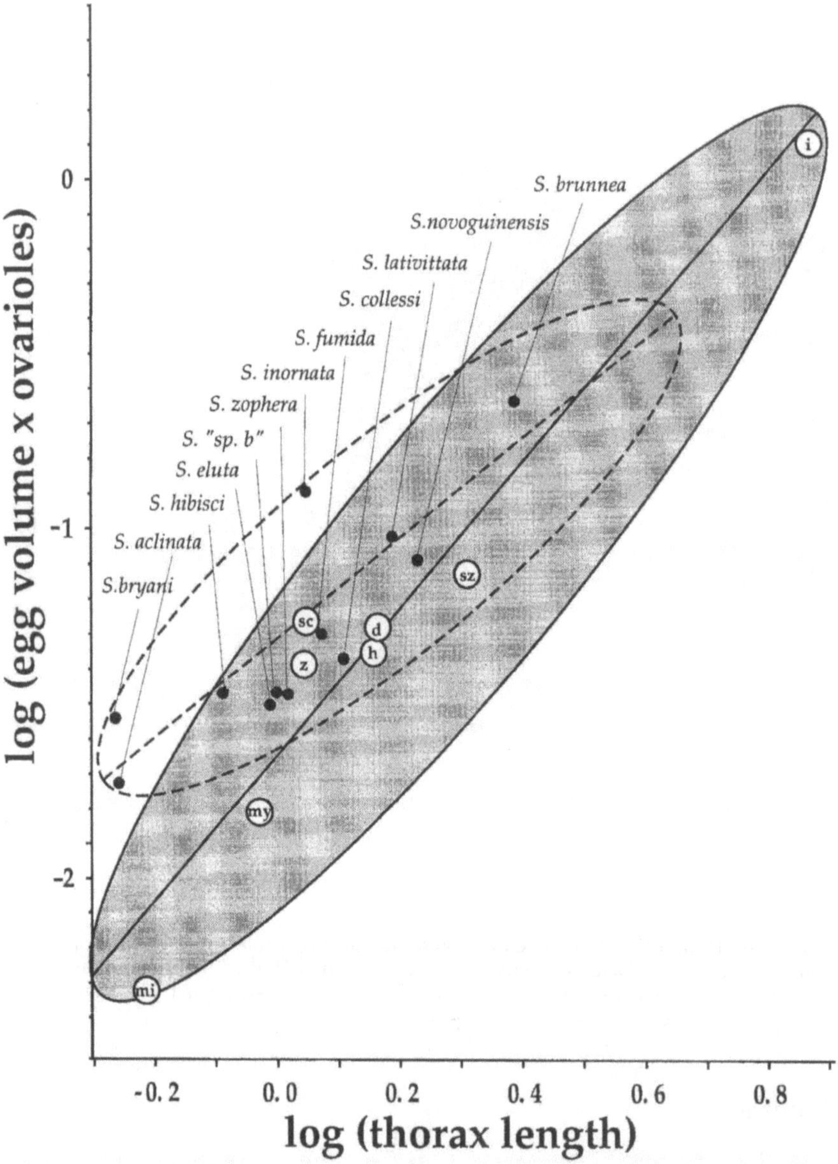

FIG. 4. Reproductive effort as a function of thorax length for genera (circles with generic abbreviations) and for species of *Scaptodrosophila* (solid circles). The extent of the family line and variation among genera is depicted as a solid line enclosed in a shaded oval. The extent of the within genus line and variation is shown as a dashed line within a dashed open oval. sz = *Scaptomyza*, h = *Hirtodrosophila*, sc = *Scaptodrosophila*, z = *Zygothrica*, mi = *Microdrosophila*, my = *Mycodrosophila*, d = *Drosophila*, and i = *Idiomyia*.

The relationship between the inter- and intraspecific pattern is of interest because the microevolutionary process is expected to translate into macro-evolutionary patterns (Hansen and Martins, 1996). A comparison of β_0 for *S. aclinata* and *S. hibisci* with the β_0 estimated from *Scaptodrosophila* species means shows that even though the intraspecific slopes are generally > -1, their 95% confidence intervals overlap with the confidence interval of the interspecific estimates of the parameter. Figures 2 and 3 compare four levels in the taxonomic hierarchy (Family > Genus > Species > Population) for the trade-off between relative egg size and ovariole number. The increased variation and general lack of fit to Model 3 is apparent at the species level but not above or below.

One problem with the comparison of spatial and temporal variation of β_0 for *S. hibisci* to higher levels of divergence is that this species is on the upper extreme of egg size for the *Scaptodrosophila* species we examined (Fig. 2). Thus, *S. hibisci* could represent a species under pressure to maintain a large egg of constant size, resulting in $\beta_0 > -1$. Unfortunately, no other *Scaptodrosophila* species had an adequate sample size to make a meaningful comparison at intermediate points on the reproductive continuum. Eleven of the North American species had sample sizes of eight or more. These species were used to compare the CV for egg volume with mean thorax length and mean ovariole number. The expected quadratic relationship of reduced variation in egg size at the extremes was not supported for either comparison. This result is not consistent with the notion that variation in egg size is constrained at extremes of the distribution. A similar analysis with the CVs for egg length and egg width of the Hawaiian species resulted in the same conclusion.

Allometry of Reproductive Allocation (ev × ova) and Thorax Length

Expression (1) estimates reproductive effort as the ratio of two volumes and assumes thorax length cubed approximates fly volume. Under this assumption changes in reproductive effort should reflect changes in allocation to locomotion, maintenance, longevity or other non-reproductive activities of the female. Model 2 does not make this assumption but estimates the scaling relationship (β_1) that would maintain a constant reproductive effort.

Table 3 indicates that the slopes of the regression of log (**ev** × **ova**) on log (**tl**) are homogenous for all levels of comparison except for the two species (*S. aclinata* versus *S. hibisci*). In general, the slopes were closer to 2 than to 3 (Table 5). The goodness of fit (captured in r^2, Table 5) was poor for *S. hibisci* but fairly good at higher and lower levels of taxonomic organization.

Model 4 evaluates egg size as a function of both egg number (**ova**) and fly size (**tl**). The β_1 estimates were homogeneous for each level of comparison (Table 3). Table 5 shows estimates of β_1 from Model 2 and Model 4 were not significantly different for regional, intergeneric (family) and generic comparisons in each analysis. In addition, improvement in fit from the two parameter models (2 and 3) to the three parameter model (4) was generally non-significant and thus the two parameter models (2 and 3) are adequate to estimate the relationships. The species data showed poor fits to Model 4 and the isofemale line analysis was mixed (Table 5).

Evolution of Reproductive Effort

The general conclusion of the reproductive effort analysis is that bigger females allocate proportionally less of their resources into reproductive tissue. There are several possible explanations for this result.

Ecological: If oviposition opportunities are rare with relatively long spans of waiting, then selection will favor greater longevity. This can be achieved in a number of ways, one of which is to favor larger animals with greater investment into maintenance at the expense of reproductive tissue. Kambysellis et al. (1995) describe this reproductive strategy for bark-breeding Hawaiian picture-winged species (*Idiomyia*). These species have rare oviposition opportunities and large bodies. They have many ovarioles and make many eggs per ovariole for each oviposition opportunity. These species are at the low end of the reproductive continuum with relatively low reproductive effort relative to body size. They also occupy the "many small eggs" corner of the trade-off expectation (Fig. 1). Small eggs may mean that the habitat imposes little pressure on larval development time. Bark-, flux- and large fruit-breeding habitats fit this criterion. The larval habitats of most of the *Scaptodrosophila* are not well known but the reproductive effort of *S. brunnea* [found on the flux of *Frareiseedendron laurifolium* (Sterculiaceae)] is consistent with the Hawaiian model.

Furthermore, when Schwarzkopf et al. (1999) applied selection for small eggs in *D. melanogaster*, the reproductive allocation decreased. This result parallels our finding that large flies with many small eggs exhibit reduced reproductive effort. However, the assumption that each ovariole has one egg is important to consider, especially if there are deviations from this pattern as mentioned above. Kambysellis and Heed (1971) report the number of eggs per ovariole in field caught females in the Hawaiian species. Two statistics are derivable from their data: 1) a positive correlation between ovariole number and eggs per ovariole ($r = 0.710, n = 36, P < 0.001$), and 2) the estimate of β_1 in Model 3 is close to the cube rather than the square ($\hat{\beta}_1 = 3.259 \pm 0.315$, $r^2 = 0.759$). Thus, when the positive relationship between egg number per ovariole and ovariole number is considered, reproductive effort is close to constant in the Hawaiian taxa. This type of analysis was not possible for the regions because either the flies assayed were from laboratory cultures or the necessary observations on eggs per ovariole were not made.

If oviposition opportunity is common but the larval habitat constrains the juvenile stage (e.g., poor nutrients or a time limit on nutrient availability) then well provisioned eggs that enhance larval competitive ability or growth rate should be favored. This condition may require more reproductive effort and, therefore, less material devoted to maintenance of the adult. The Hawaiian taxa that live as larvae under these conditions are the leaf-breeders and flower-breeders. These flies are on the higher end of the reproductive effort continuum and also occupy the "few-large eggs" corner of the trade-off expectation. It is noteworthy that all three geographic regions have taxa that exemplify this strategy (few-large eggs) and in all cases they are flower-breeders.

Morphological: The $\beta_1 \approxeq 2$ result (i. e., the reproductive effort scales to the square of thorax length and not to the cube) may be explained by the fact that ovarioles and eggs are in the expandable abdomen rather than a fixed space such as the thorax. It is thus possible that increasing a linear dimension (thorax length) increases the capacity for expansion of the abdomen such that it does not require as large an increase in egg size × egg number to achieve the same proportional reproductive effort. However, even after accounting for this (Model 2 and 4), residual differences in reproductive effort still existed for different genera of the family and for the two closely related species of *Scaptodrosophila* (Table 3). In this case, the size factor was not important. Genera with larger flies exhibited relatively high and low effort (*Idiomyia* and *Scaptomyza*, respectively), as did genera with smaller flies (*Scaptodrosophila* and *Mycodrosophila* had relatively high and low effort, respectively) (Fig. 4). The only significant correlation with residual reproductive effort was with mean ovariole number.

In addition, there may be a physical space trade-off in the abdomen if crop volume (which can ultimately go to either maintenance or reproduction) competes with ovarian volume. Even though the crop is anterior to the ovaries, when either increases in volume, space for the other may become restricted.

The linear relationship between log of reproductive volume and log thorax length is consistent in the phylogenetic hierarchy (Family > Genera > Species > Population > Individual). The range covered by the relationship is decreasing until the level of population, i. e., Family > Genus > Species > Population < Individual. The analysis of the isofemale lines of *S. hibisci* did not show significant heritability with most of the variance within lines (89%) and little (1%) for lines within populations. This result is illustrated in Figs. 4, 5 and 6. The increased variance from isofemale lines to females (within line) illustrates the large environmental component that influences reproductive effort and the resulting phenotypic plasticity.

The extraordinary range of egg size and ovariole number in the family Drosophilidae appears to have occurred independently in several regions

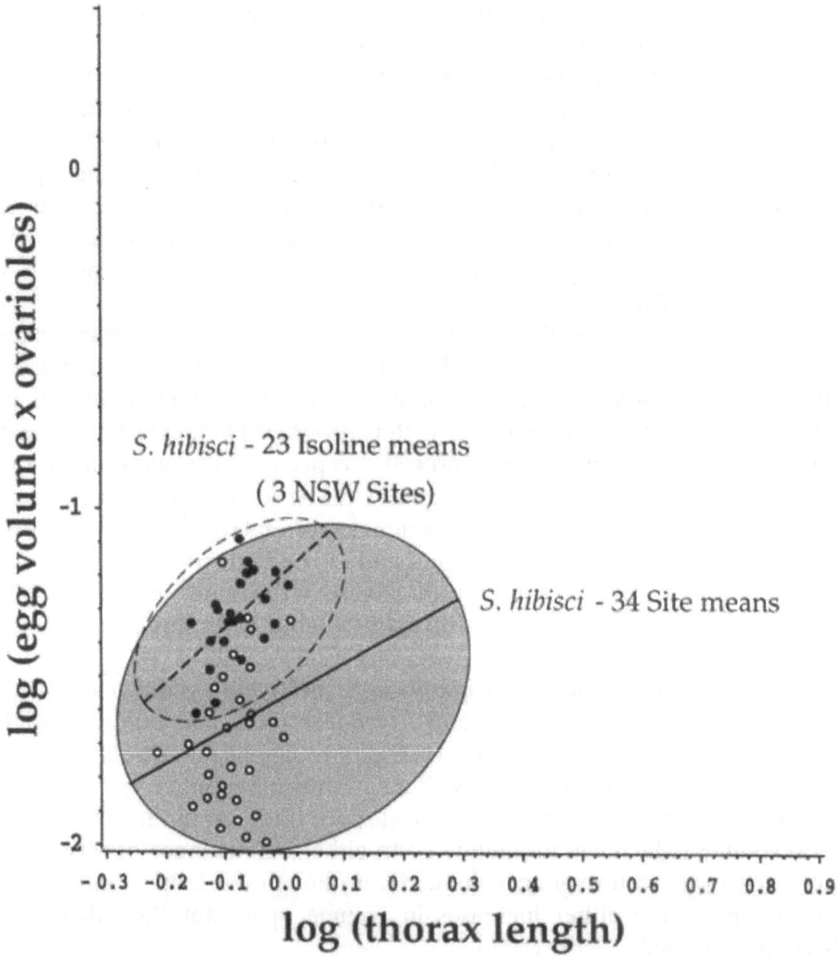

FIG. 5. Reproductive effort as a function of thorax length for population means (sites) of *S. hibisci* in eastern Australia (open circles with shaded oval boundary) and for isoline means of three sites in N. S. W. (closed circles with dashed line and dashed oval boundary).

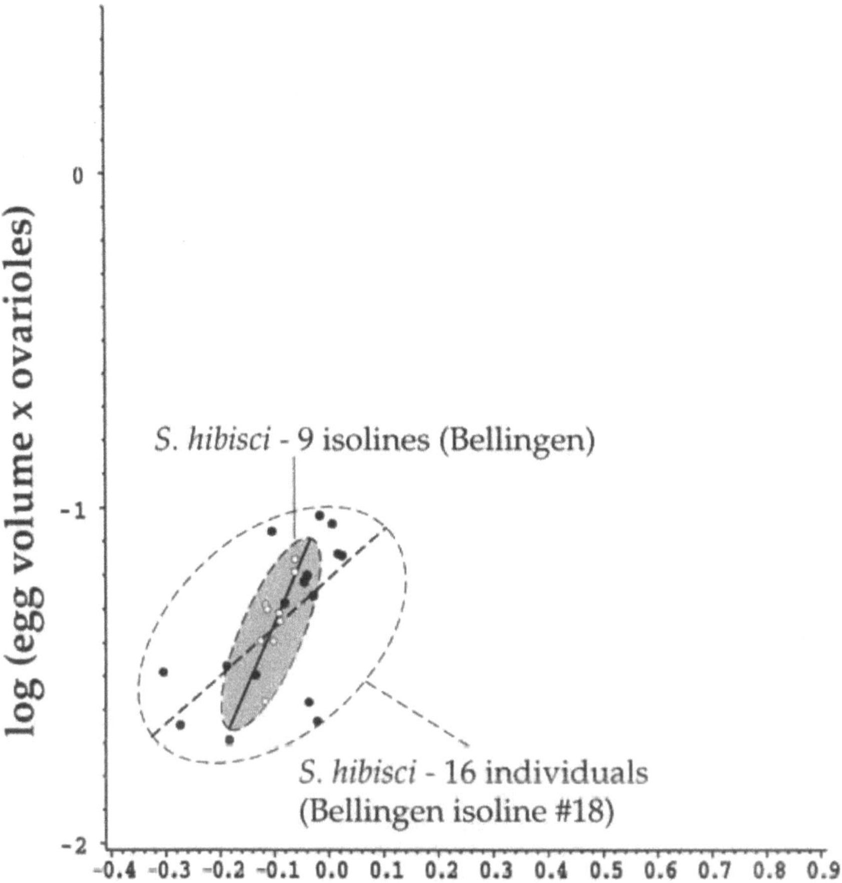

FIG. 6. Reproductive effort as a function of thorax length for isofemale lines of *S. hibisci* collected from Bellingen, N. S. W. (open circles with shaded oval boundary) and for individuals of a representative isoline (closed circles with dashed line and dashed oval boundary).

of the world. The negative relationship between egg size and number is also common and is manifested at all levels of the taxonomic hierarchy. Even though the relationship is robust to violation of several assumptions, deviations from isometry are worthy of further study. Lifetime fecundity studies, larval habitat discovery and more precise body metrics for relating body mass to reproductive effort would contribute to a better understanding of reproductive trade-offs in these flies.

ACKNOWLEDGMENTS

This work was supported by a grant from International Programs at NSF.

References

Ashburner, M., 1989, *Drosophila*: a laboratory handbook. Cold Spring Harbor Laboratory Press, New York.

Atkinson, W. D., 1979, A comparison of reproductive strategies of domestic species of *Drosophila*. *J. Anim. Ecol.* **48:**53–64.

Avelar, T., 1993, Egg size in *Drosophila*: standard unit of investment or variable response to environment? The effect of temperature. *J. Insect Physiol.* **39:**283–289.

Avelar, T., and Rocha Pité, M. T., 1989, Egg size and number in *Drosophila subobscura* under semi-natural conditions. *Evol. Biol.* **3:**37–48.

Bernardo, J., 1996, The particular maternal effect of propagule size, especially egg size: patterns, models, quality of evidence and interpretation. *Am. Zool.* **36:**216–236.

Berrigan, D., 1991, The allometry of egg size and number in insects. *Oikos* **60:**313–321.

Beverley, S. M., and Wilson, A. C., 1982, Molecular evolution in *Drosophila* and the higher Diptera. I. Micro-complement fixation studies of a larval hemolymph protein. *J. Mol. Evol.* **18:**251–264.

Beverley, S. M., and Wilson, A. C., 1984, Molecular evolution in *Drosophila* and the higher Diptera. II. A time scale for fly evolution. *J. Mol. Evol.* **21:**1–13.

Bock, I. R., 1976, Drosophilidae of Australia I. *Drosophila* (Inseceta: Diptera). Australian Journal of Zoology. *Suppl. Ser.* **40:**1–105.

Boulétreau-Merle, J., Allemand, R., Cohet, Y., and David, J. R., 1982, Reproductive strategy in *Drosophila melanogaster*: significance of a genetic divergence between temperate and tropical populations. *Oecologia* **53:**323–329.

Brockelman, W. Y., 1975, Competition, the fitness of offspring, and optimal clutch size. *Am. Nat.* **109:**677–699.

Caccone, A., Gleason, J. M., and Powell, J. R., 1992, Complementary DNA-DNA hybridization in *Drosophila*. *J. Mol. Evol.* **34:**130–140.

Carlson, K. A., Nusbaum, T. J., Rose, M. R., and Harshman, L. G., 1998, Oocyte maturation and ovariole number in lines of *Drosophila melanogaster* selected for postponed senescence. *Functional Ecology* **12:**514–520.

Carson, H. L., and Kaneshiro, K. Y., 1976, *Drosophila* of Hawaii: systematics and ecological genetics. *Ann. Rev. Ecol. Syst.* **7:**311–346.

Carson, H. L., and Okada, T., 1982, Drosophilidae of New Guinea. pp. 675–687, in: Biogeography and ecology of New Guinea (J. L. Gressitt, ed., *Monogr. Bio. Vol.* **42(2)** W. Junk, The Hague.

Chen, P. S., Stumm-Zollinger, E., Aigaki, T., Balmer, J., Bienz, M., and Bohlen, P., 1988, A male accessory gland peptide that regulates reproductive behavior of female *D. melanogaster*. *Cell* **54:**291–298.

Cody, M. L., 1966, A general theory of clutch size. *Evolution* **20:**174–184.

Darwin, C., 1872, The Origin of Species. Mentor Edition, Penquin Inc., NY.

David, J. R., 1970, Le nombre d'ovarioles chez la *Drosophila*: relation avec la fecondite et valeur adaptive. *Arch. Zool. Exp. Gén.* **111:**357–370.

DeSalle, R., 1992, The phylogenetic relationships of flies in the family Drosophilidae deduced from mtDNA sequences. *Mol. Phylog. Evol.* **1:**31.

Eberhard, W. G., 1996, Female control: sexual selection by cryptic female choice. Princeton, NJ: University Press.

Felsenstein, J., 1985, Phylogenies and the comparative method. *Am. Nat.* **125:**1–15.

Gasser, M., Kaiser, M., Berrigan, D., and Stearns, S. C., 2000, Life history correlates of evolution under high and low adult mortality. *Evolution* **54:**1260–1272.

Grimaldi, D. A., 1990, A phylogenetic, revised classification of genera in the Drosophilidae (Diptera). Bulletin of the American Museum of Natural History. **197:**1–139.

Grimaldi, D., James, A. C., and Jaenike, J., 1992, Systematics and modes of reproductive isolation in the holarctic *Drosophila testacea* species group (Diptera: Drosophildae). *Ann. Entomol. Soc. Am.* **85:**671–685.

Hansen, T. F., and Martins, E. P., 1996, Translation between microevolutionary process and macroevolutionary patterns: The correlation structure of interspecific data. *Evolution* **50:**1404–1417.

Harvey, P. H., and Pagel, M. D., 1991, The comparative method in evolutionary biology. Oxford University Press, Oxford.

Herndon, L. A., and Wolfner, M. F., 1995, A *Drosophila* seminal fluid protein, Acp26Aa, stimulates egg-laying in females for one day following mating. *Proc. Natl. Acad. Sci. USA* **92:**10114–10118.

Hirshfield, M. F., and Tinkle, D. W., 1975, Natural selection and the evolution of reproductive effort. *Proc. Natl. Acad. Sci. USA* **72:**2227–2231.

Holland, B., and Rice, W. R., 1998, Chase-away sexual selection: antagonistic seduction versus resistance. *Evolution* **52:**1–7.

Kalb, J. M., DiBenedetto, A. J., and Wolfner, M. F., 1993, Probing the function of *Drosophila melanogaster* accessory glands by directed cell ablation. *Proc. Natl. Acad. Sci. USA* **90:**8093–8097

Kambysellis, M. P., and Heed, W. B., 1971, Studies of oogenesis in natural populations of Drosophilidae. I. Relation of ovarian development and ecological habitats of the Hawaiian species. *Am. Nat.* **105:**31–49.

Kambysellis, M. P., Starmer, W. T., Smathers, G., and Heed, W. B., 1979, Studies of oogenesis in natural populations of Drosophilidae. II. Significance of microclimatic changes on oogenesis of *D. mimica*. *Am. Nat.* **115(1):**67–91.

Kambysellis, M. P., Ho, K.-F., Craddock, R. M., Piano, F., Parisi, M., and Cohen, J., 1995, Pattern of ecological shifts in the diversification of Hawaiian *Drosophila* inferred from a molecular phylogeny. *Current Biology* **5:**1129–1139.

Kermack, K. A., and Haldane, J. B. S., 1951, Organic correlation and allometry. *Biometrika* **37:**30.

King, R. C., 1970, Ovarian development in *Drosophila melanogaster*. Academic Press, New York.

King, R. C., and Sang, J. H., 1959, Oogenesis in adult *Drosophila melanogaster*. VIII. The role of folic acid in oogenesis. *Growth* **23:**37–53.

Kwiatowski, J., Skarecky, D., Bailey, K., and Ayala, F. J., 1994, Phylogeny of *Drosophila* and related genera inferred from the nucleotide sequence of the Cu, Zn *Sod* gene. *J. Mol. Evol.* **38:**443–454.

Lack, D., 1947, The significance of clutch size. *Ibis* **89:**302–352.

Lloyd, D. G., 1987, Selection of offspring size and other size-versus-number strategies. *Am. Nat.* **129:**800–817.

McEvey, S. F., and Barker, J. S. F., 2001, *Scaptodrosophila aclinata*: A new Hibiscus flower-breeding species related to *S. hibisci* (Diptera: Drosophilidae). *Rec. Aust. Mus.* **53:**255–262.
McGinley, M. A., Temme, D. H., and Geber, M. A., 1987, Parental investment in offspring in variable environments: theoretical and empirical considerations. *Am. Nat.* **118:**370–398.
Montague, J. R., 1984, The ecology of Hawaiian flower-breeding drosophilids. 1. Selection in the larval habitat. *Am. Nat.* **124:**712–722.
Montague, J. R., 1989, The ecology of Hawaiian flower-breeding drosophilids. 2. Adult dispersions and reproductive ecology. *Am. Nat.* **133:**71–82.
Montague, J. R., Mangan, R. L., and Starmer, W. T., 1981, Reproductive allocation in the Hawaiian Drosophilidae: egg size and number. *Am. Nat.* **118:**865–871.
Parker, G. A., and Begon, M., 1986, Optimal egg size and clutch size: effects of environment and maternal phenotype. *Am. Nat.* **128:**573–592.
Parsons, P. A., and Bock, I. R., 1979, The population biology of Australian *Drosophila*. *Ann. Rev. Ecol. Syst.* **10:**229–245.
Pelandakis, M., and Solignac, M., 1993, Molecular phylogeny of *Drosophila* based on ribosomal RNA sequences. *J. Mol. Evol.* **37:**525–543.
Pitnick, S., and Markow, T. A., 1994, Large-male advantages associated with costs of sperm production in *Drosophila hydei*, a species with giant sperm. *Proc. Natl. Acad. Sci. USA* **91:**9277–9281.
Pitnick, S., Markow, T. A., and Spicer, G. S., 1999, Evolution of multiple kinds of female sperm-storage organs in *Drosophila*. *Evolution* **53:**1804–1822.
Powell, J. R., 1997, Progress and prospects in evolutionary biology: the *Drosophila* model. Oxford University Press, New York.
Powell, J. R., and DeSalle, R., 1995, *Drosophila* molecular phylogenies and their uses. *Evol. Biol.* **28:**87–138.
Purvis, A., and Rambaut, A., 1995, Comparative analysis by independent contrasts (CAIC): an Apple Macintosh application for analyzing comparative data. *Comp. Appl. Biosci.* **11:**247–251.
Rayner, J. M. V., 1985, Linear relations in biomechanics: the statistics of scaling functions. *J. Zool. Lond (A)* **206:**415–439.
Rice, W. R., 1998, Intergenomic conflict, interlocus antagonistic evolution, and the evolution of reproductive isolation, in: Endless forms: species and speciation (D. J. Howard and S. H. Berlocher, ed.) pp. 261–270. Oxford University Press, Oxford.
Robertson, F. W., and Reeve, E. C. R., 1953, Studies in quantitative inheritance IV. The effects of substituting chromosomes from selected strains in different genetic backgrounds in *Drosophila melanogaster*. *Journal of Genetics* **51:**586–610.
Russo, C. A. M., Takezaki, N., and Nei, M., 1995, Molecular phylogeny and divergence times of Drosophilid species. *Mol. Biol. Evol.* **12:**391–404.
Sargent, R. C., Taylor, P. D., and Gross, M. R., 1987, Parental care and the evolution of egg size in fishes. *Am. Nat.* **129:**32–46.
Schwarzkopf, L., Blows, M. W., and Caley, M. J., 1999, Life-history consequences of divergent selection on egg size in *Drosophila melanogaster*. *Am. Nat.* **154:**333–340.
Sinervo, B., Doughty, P., Huey, R. B., and Zamudio, K., 1992, Allometric engineering: a causal analysis of natural selection on offspring size. *Science* (Washington, D. C.) **258:**1927–1930.
Smith, C. C., and Fretwell, S. D., 1974, The optimal balance between size and number of offspring. *Am. Nat.* **108:**499–506.
Spicer, G. S., 1988, Molecular evolution among some *Drosophila* species groups as indicated by two-dimensional electrophoresis. *J. Mol. Evol.* **27:**250–260.

Spicer, G. S., 1991, Molecular evolution and phylogeny of the *Drosophila virilis* species group as inferred by two-dimensional electrophoresis. *J. Mol. Evol.* **1991:**379–394.

Spicer, G. S., 1992, Reevaluation of the phylogeny of the Drosophila virilis species group (Diptera: Drosophilidae). *Ann. Entomol. Soc. Am.* **85:**11–25.

Spicer, G. S., and Jaenike, J., 1996, Phylogenetic analysis of breeding site use and α-amanitin tolerance within the *Drosophila quinaria* species group. *Evolution* **50:**2328–2337.

Spicer, G. S., and Pitnick, S., 1996, Molecular systematics of the *Drosophila hydei* subgroup as inferred from mitochondrial DNA sequences. *J. Mol. Evol.* **43:**281–286.

Starmer, W. T., Wolf, L. L., Barker, J. S. F., Bowles, J. M., and Lachance, M. A., 1997, Reproductive characteristics of the flower breeding *Drosophila hibisci* Bock (Drosophilidae) along a latitudinal gradient in eastern Australia: relation to flower and habitat features. *Biol. J. Linn. Soc.* **62:**459–473.

Starmer, W. T., Polak, M., Wolf, L. L., and Barker, J. S. F., 1998, Reproductive characteristics of the flower breeding *Drosophila hibisci* Bock (Drosophilidae) along a latitudinal gradient in eastern Australia: genetic and environmental determinants of ovariole number. *Evolution* **52:**806–815.

Starmer, W. T., Polak, M., Wolf, L. L., and Barker, J. S. F., 2000, Reproductive characteristics of the flower breeding *Drosophila hibisci* Bock (Drosophilidae) along a latitudinal gradient in eastern Australia: within-population genetic detrminants of ovariole number. *Heredity* **84:**90–96.

Stearns, S. C., 1976, Life-history tactics: a review of ideas. *Q. Rev. Biol.* **51:**3–47.

Stearns, S. C., 1977, The evolution of life history traits: a critique of the theory and a review of the data. *Ann. Rev. Ecol. Syst.* **8:**145–171.

Stearns, S. C., 1992, The evolution of life histories. Oxford University Press, Oxford.

Sullivan, D. T., Atkinson, P. W., and Starmer, W. T., 1990, Molecular evolution of the alcohol dehydrogenase genes in the genus *Drosophila*. *Evol. Biol.* **24:**107–147.

Throckmorton, L. H., 1962, The problem of phylogeny in the genus *Drosophila*. *Univ. Texas Publ.* **6205:**207–343.

Throckmorton, L. H., 1975, The phylogeny, ecology and geography of *Drosophila*. pp. 421–469, in: Handbook of Genetics, vol. 3 (R. C. King, ed.) Plenum Press, New York.

Vance, R. R., 1973a, On reproductive strategies in marine benthic invertebrates. *Am. Nat.* **107:**339–352.

Vance, R. R., 1973b, More on reproductive strategies in marine benthic invertebrates. *Am. Nat.* **107:**353–361.

Wasserman, M., 1992, Cytological evolution of the *Drosophila repleta* species group. pp. 455–552, in: *Drosophila* inversion polymorphism (C. B. Krimbas and J. R. Powell, eds.) CRC Press, Boca Raton.

Wayne, M. L., Hackett, J. B., and Mackay, T. F. C., 1997, Quantitative genetics of ovariole number in *Drosophila melanogaster*. I. Segregating variation and fitness. *Evolution* **51:**1156–1163.

Wiklund, C., Karlsson, B., and Forsberg, J., 1987, Adaptive versus constraint explanations for egg-to-body size relationships in two butterfly families. *Am. Nat.* **130:**828–838.

Winkler, D. W., and Wallin, K., 1987, Offspring size and number: a life history model linking effort per offspring and total effort. *Am. Nat.* **129:**708–720.

Wolf, L. L., Polak, M., Barker, J. S. F., Bowles, J., and Starmer, W. T., 2000, Reproductive characteristics of *Drosophila hibisci* in the Northern Territory, Australia. *Biol. J. Linn. Soc.* **71:**549–562.

4

Benefits of Polyandry: A Life History Perspective

D. J. HOSKEN and P. STOCKLEY

INTRODUCTION

Bateman's principle has been widely interpreted to imply that females gain no fitness benefits from polyandry (Bateman, 1948) and, therefore, should not be expected to mate multiply (here defined as mating with more than one male). Nevertheless, it is increasingly clear that females of many, if not most, taxa do copulate with multiple males (e.g. Birkhead and Møller, 1998). Moreover, polyandry is widespread despite considerable costs, including wasted time and energy, increased risk of predation and disease, potential damage caused by male seminal fluids and copulatory organs, and even death (Keller and Reeve, 1995; Eberhard, 1996). Despite these associated costs, females of diverse taxa not only accept several mates but also actively solicit multiple copulations in many instances (Birkhead and Møller, 1998). As evidence of diverse potential benefits associated with polyandry now accumulates, the assumption that females should not mate multiply because they cannot increase offspring numbers by doing so appears questionable. Importantly, since females have greater potential than males to influence the quality of their offspring, and investment in current reproduction has consequences for future reproductive attempts,

D. J. HOSKEN • Zoologisches Museum, University of Zurich-Irchel, Winterthurerstrasse 190, CH-8057 Zurich, Switzerland. E-mail: hosken@zoolmus.unizh.ch, Tel: +41-1-6354973, Fax: +41-1-6356818. P. STOCKLEY • Animal Behaviour Group, Faculty of Veterinary Science, University of Liverpool, Leahurst, Neston, South Wirral CH64 4BP. E-mail: stockley@liv.ac.uk, Tel: 0151 794 6009, Fax: 0151 794 6107.

they should be expected to optimise offspring numbers rather than maximise numbers produced in any given reproductive attempt (Roff, 1992; Stearns, 1992). Moreover, the relationship between female lifetime reproductive success and offspring numbers and/or size may not be as straightforward as is generally assumed (e.g. Madsen and Shine, 1998; Stockley and Macdonald, 1998). To suggest that females should not mate multiply simply because they cannot increase offspring numbers in a given reproductive attempt by doing so is therefore to confuse the currency of male and female fitness, and thereby to considerably under-estimate the complexity of female reproductive strategies.

While evidence is rapidly accumulating for diverse potential benefits of female multiple mating behaviour, the extent to which general hypotheses might apply across taxa is still unclear. Further, recent advances in understanding the evolutionary causes and consequences of sexual conflict suggest that an important distinction should be made between the type of net benefits accruing to females i) when polyandry is initiated primarily by males and ii) when it is also initiated by females. In the first case, females are reluctant to remate in the absence of male persistence or coercion, suggesting that benefits of polyandry are most likely to be associated with cost minimisation. In the second, when females actively seek to remate in many instances, costs of mating are apparently outweighed by specific benefits relating to female longevity (and hence number of reproductive attempts), clutch size and/or offspring quality. The aim of this review is to consider examples of benefits of both male- and female-initiated polyandry in relation to major life history components, and to highlight potential generalisations emerging from this treatment.

MALE INITIATED POLYANDRY: BENEFITS RESULTING FROM COST MINIMIZATION

Resource and energy availability is a key determinant of female reproductive success at each life history stage from timing of sexual maturation to overall longevity across diverse taxa (Emlen and Oring, 1977 and see e.g. mammals, Festa Bianchet et al., 1998). Consequently, female reproductive success is commonly influenced by factors that constrain or improve rates of energy gain.

Harassment by courting males may represent a significant constraint on the amount of time that a female can spend foraging, and/or a significant drain on energy stores if she must expend valuable energy in rejecting male advances. Cost minimisation may drive female multiple mating if it is cheaper (in terms of time and/or energy) for females to accept additional copulations than to try and resist male advances (Parker, 1979). Clutton-Brock and Parker (1995) collate convincing evidence from a wide range of taxa showing that females accept copulations from harassing males when they cannot evade them. In such species, males typically force copulation, punish females which will not copulate with them or harass them until they copulate. Waterstriders are a model example of female multiple mating as a means of cost minimisation (i.e. convenience polyandry). The predominant mating pattern in this group involves intense pre- and post-copulatory struggles between males and females (Arnqvist, 1997). In addition to the energetic costs of such struggles, there are often additional costs associated with copula, including increased predation and injury risk (e.g. Arnqvist, 1989; Fairbain, 1993). In spite of these obvious costs to females, there are no obvious benefits to balance them (e.g. nutrient donation, good genes, sperm replenishment; but see Lauer, 1996), although for females paired with a male, harassment is reduced and feeding efficiency may increase. Nevertheless, female foraging rewards would probably be greater if they could avoid males altogether but, in all species studied to date, females copulate frequently (Arnqvist, 1997). Thus, in many waterstriders, the evidence is consistent with female multiple mating as a means of minimising resistance costs. Moreover, in at least one species, sexual conflict over mating has selected for morphological adaptations that help females resist unwanted copulations (Arnqvist and Rowe, 1995).

FEMALE INITIATED POLYANDRY: MATERIAL BENEFITS

A recent review of polyandry in insects showed females from nuptial feeding species achieved increasing lifetime offspring production from polyandry, and the authors concluded genetic benefits were not necessary for the maintenance of polyandry in this taxon (Arnqvist and Nilsson, 2000). Although some studies included in that meta-analysis may have confounded genetic and other benefits (i.e. females not only mated multiply, they mated to more than one male), material benefits to females are likely to be important and can influence a number of female life history components.

1. Benefits Influencing Female Lifespan

Longevity is also an important determinant of lifetime reproductive success across diverse taxa (Clutton-Brock, 1988). This is true even when fertility declines with age. Two main kinds of benefits have been associated with polyandry in relation to female lifespan, each of which effectively maximises opportunities to increase the total number of reproductive attempts. The first acts by increasing female longevity via the provision of nutritional benefits, which apparently function to reduce costs of reproduction, while the second reduces female risk of mortality via extrinsic causes such as predation.

i) Nutritional Benefits

Males in a variety of taxa offer incentives to encourage females to copulate. These include energetic and other benefits which can influence female condition and hence lifetime reproductive success. For example Burpee and Sakaluk's (1993) study showed that female longevity increased with daily spermatophore consumption in several crickets (see also e.g. Wedell, 1996). Similarly, Karlsson (1998) described increased female longevity (and also egg number) with increasing spermatophore number in a polyandrous butterfly, and Fox (1993) showed that under conditions of nutrient stress, ejaculate derived nutrients contribute to female somatic maintenance, increasing female longevity in bruchid beetles (but see Savalli and Fox, 1999 for the opposite effect in another population). More recently, Brown (1997) has shown that nuptial feeding increased female lifespan and hence lifetime fecundity in a tree cricket. Importantly, this effect was independent of oviposition rate.

ii) Reduced Predation Risk

Male moths (*Utetheisa ornatrix*) transfer defensive alkaloids to females during copula, which provide protection from spider, and possibly other, predators (Gonzáles et al., 1999). However, although females gain prolonged protection from a single copulation, they typically mate 4–5 times over their life-time and show a preference for males with high alkaloid content, suggesting that alkaloids are an important factor driving female multiple mating. A similar combination of benefits has also been demonstrated for female scorpion flies. Here females spend less time flying and foraging as a result of courtship feeding by males and this greatly reduces predation risk: about twice as many males as females are killed in spider webs (Thornhill, 1976, 1978).

2. Benefits Via Offspring Number

In contrast to Bateman's (1948) study, females of certain taxa do apparently benefit from polyandry by increasing offspring numbers (Arnqvist and Nilsson, 2000). This can occur via simple fertility/fecundity increases, infanticide avoidance and/or the provision of additional parental investment.

i) Fertility

At the most basic level, a female's reproductive success is dependent on fertilisation of the ova she produces. Polyandry may therefore function simply to promote fertility benefits, as may occur when females cannot store sufficient sperm to last their reproductive lifetime. Sexual conflict over ejaculate size may also select for female multiple mating. For example, in external fertilisers such as many fishes, there is evidence that female infertility can result from male optimisation of ejaculate expenditure under certain conditions (Ball and Parker, 1996; Warner et al., 1995; see also *Drosophila*, Pitnick and Markow, 1994). Under such circumstances, females should be selected to spawn simultaneously with several males, and thereby reduce the level of infertility incurred (Ball and Parker, 1996). Similarly, Boomsma (1996) suggested that female multiple mating in hymenopterans may be the result of male ejaculation strategies, with males allocating only a small ejaculate to each mate, not filling the female sperm stores, and hence favouring polyandry.

Multiple mating may increase the number of sperm stored by females. For example, the number of sperm stored by Panamanian leaf-cutter ant queens increases by approximately 30 million with each additional mating, and mating with more than one male is probably needed to maintain fertility over a queen's reproductive lifespan (Fjerdingstad and Boomsma, 1998). Fertility benefits of this nature have been suggested as a general explanation for multiple mating in insects (Ridley, 1989), and possibly also relate to sperm degradation during storage, male sterility, or ejaculates being insufficiently large to fertilize whole clutches (Gibson and Jewell, 1982; Halliday and Arnold, 1987). Selection favouring polyandry via fertility benefits has been advocated less frequently for birds and mammals however. For example, some evidence suggestive of polyandry as fertility insurance exists in birds (e.g. Sax et al., 1998), but a number of other data suggest that fertility insurance is not the primary mechanism promoting female multiple mating across this taxon. For instance, Birkhead and Møller (1992) suggested that if infertility were a widespread problem, colonial birds should have lower incidence of infertile eggs compared with solitary species since in the former the opportunity for extra-pair copulations (EPC's) is higher. No relationship was found (Birkhead and Møller, 1992), and it has been argued that generally, male infertility is unlikely to be common enough to be a widespread mechanism favouring female multiple mating (Alcock et al., 1978; also see Gibson and Jewell, 1982; Birkhead and Møller, 1992).

ii) Fecundity

An elevated level of parental investment, leading to increased offspring production, can occur via donation of ejaculatory nutrients as shown in a range of insects including butterflies (Karlsson, 1998) and *Drosophila* (e.g. Markow et al., 1990; Pitnick et al., 1991). For example, Markow et al. (1990) showed that male derived nutrients increased egg production in *D. mojavensis* at least over the subsequent 10 days, and Gwynne (1984) also showed a positive relationship between nuptial gift consumption and the number and size of eggs produced by the bushcricket *Requena verticalis*. Similarly, in crickets, both egg number and hatchability was positively associated with number of matings (Simmons, 1988). Where males control access to resources, females may also increase offspring numbers by mating multiply when resources are scarce. Alcock (1989) called this behaviour "prostitution polyandry". The bushcricket *Kwanaphila nartee* is an excellent example of this. When resources (pollen) are scarce, individuals become sex-role reversed and females forage for male spermatophores which can allow them to maintain their reproductive output at abundant resource levels (Simmons and Bailey, 1990).

iii) Increased Paternal Care

In species with parental investment by both sexes, females may benefit from polyandry by soliciting additional help in rearing young. For example, female dunnocks (*Prunella modularis*) often copulate with two males per clutch, and both males subsequently help the female with offspring rearing (Davies, 1985; Davies et al., 1996). Similarly, in the polygynandrous Smith's longspur, females mate with one to three males, most of which subsequently assist in feeding the offspring (Briskie et al., 1998). Importantly, paternal feeding in Smith's longspur greatly increases offspring survival (unpublished data in Briskie et al., 1998). Polyandry also increases help at the nest and brood provisioning in the Alpine accentor, which in turn leads to increased nestling mass and survival (Nakamura, 1998). Thus increased paternal contribution associated with polyandry is likely to have increased optimal clutch size over evolutionary time.

iv) Infanticide Avoidance

Infanticide avoidance has been proposed as a primary reason for female multiple mating in a number of mammal species, particularly primates, rodents and some felids (e.g. Hrdy, 1977). Infanticide is adaptive for males under certain conditions where it is directed at offspring of competing males and brings females into oestrus more quickly (e.g. Borries et al., 1999). By mating multiply, females may confuse paternity and so reduce the risk of infanticide, as males will be unlikely to kill offspring which they have potentially sired. For example, Cicirello and Wolff (1990) found that male white-footed mice (*Peromyscus leucopus*) were inhibited from killing pups for a period after copulation approximating to the total duration of gestation and lactation for this species. In non-mammalian taxa the evidence is less compelling, and in some instances females are unwilling to copulate with infanticidal males (e.g. Møller, 1988), exactly the opposite relationship expected if infanticide avoidance were driving female multiple mating.

v) Predation Avoidance

In invertebrates, there is evidence for the use of male donated chemicals to protect eggs from predation and hence increase offspring number (LaMunyon, 1997; Gonzales et al., 1999). In the moth *Utetheisa ornatrix* (see II above), not only do male derived alkaloids protect females from predation, but these defence chemicals are also provided to the eggs, protecting them from attack by lacewing larvae (e.g. Elsner et al., 2000).

FEMALE INITIATED POLYANDRY: GENETIC BENEFITS

It has recently been suggested that all benefits of polyandry will eventually become associated with genetic benefits (Jennions and Petrie, 2000). However, this assumes that all benefits are either related to genetic compatibility, have a heritable basis, or correlate with some measure of mate quality which is heritable. At present the evidence for genetic compatability benefits is limited (Tregenza and Wedell, 2000 but see below). It is also possible that benefits are not heritable if environmental variance or dominance or epistatic variance are large, or genetic benefits may be environment dependent, being favoured in one environment and time but not in another (GxE interactions) (e.g. Vieira et al., 2000). Additionally, while non-genetic benefits may correlate with mate quality, this need not always occur. Yasui (1998) suggested non-genetic benefits are most likely to be (at least) the initial female benefit from polyandry, and a recent review of polyandry in insects suggested suggests that direct benefits explained explain multiple mating in this group (Arnqvist and Nilsson, 2000). Nevertheless, substantial evidence exists suggesting that polyandry provides genetic benefits to females in a range of taxa (e.g. Jennions and Petrie, 2000). Benefits may accrue via overt or cryptic mechanisms, and may involve direct or indirect female choice (cf. Wiley and Poston, 1996), but need not always result in directional sexual selection.

i) Mate Choice

It has often been suggested that females of socially monogamous bird species mate multiply to compensate for having a mate of low genetic quality (Birkhead and Møller, 1992). Multiple mating may thus allow females to increase the quality of their offspring via heritable genetic benefits gained from extra-pair copulations with high quality males. For example, Petrie et al. (1998) found a positive relationship between population genetic diversity, as measured by enzyme and RAPD heterozygosity, and EPC's. This is consistent with mate choice for good genes, as in more homozygous populations females have little to gain (all males are about the same) and hence are more monogamous. However, in more heterozygous populations the ability to detect multiple paternity is also greater.

ii) Sperm Competition

Several kinds of potential genetic benefits have been suggested to result from female promotion of sperm competition. For example, Birkhead et al. (1993) suggested that females might gain genetic benefits for their offspring following multiple mating if the fittest males are generally most successful at fertilising their eggs during sperm competition. This kind of 'intrinsic male quality' hypothesis often assumes haploid expression, for which there is some evidence (Erickson, 1990). Good genes benefits were suggested initially by Madsen et al. (1992) to explain their finding that female adders (*Vipera berus*) mating with multiple males produced more viable offspring. They argued that this advantage resulted from more intense sperm competition in the female's reproductive tract, resulting in a higher proportion of her ova being fertilized by genetically superior males. However, subsequent studies (Olsson et al., 1996) suggest that results of the adder study might be better interpreted in terms of selection for genetic compatibility rather than success in sperm competition of genetically superior males. An alternative potential mechanism by which the intrinsic male quality hypothesis could work is if the 'fittest' males are generally the most heterozygous (Brown, 1997). This could potentially select for female multiple mating because there may be no reliable indicator of heterozygosity *per se* (e.g. the relationship between heterozygosity and fluctuating asymmetry, proposed as one means of recognising heterozygosity, appears very weak, and is absent in many taxa; Vøllestad et al., 1999; and see Hosken et al., 2000). If males with low heterozygosity have relatively poor quality sperm such a mechanism could apply, although there is limited evidence for this (e.g. Wildt et al., 1987). Nevertheless, heterozygosity benefits will typically only accrue when sperm selection (of some form) occurs (Partridge, 1983, and see below). However, the only study to date which explicitly investigates the heterozygosity hypothesis in a natural population found no relationship between heterozygosity levels and reproductive success (Weatherhead et al., 1999).

Further potential for genetic benefits of multiple mating is suggested by the sexy sperm hypothesis (Keller and Reeve, 1995). Here, it is suggested that females benefit by multiply mating since by doing so they produce sons which do well in sperm competition. A key assumption of this hypothesis is that sperm competition ability is heritable. In support of this idea, Radwan (1998) found some evidence in the bulb mite, *Rhizoglyphus robini,* that male success in sperm competition has a heritable component (see also Hosken et al., 2001). Although the sexy sperm hypothesis could potentially apply across diverse taxa, there is currently limited evidence for it in most groups. Nevertheless, there is evidence that high quality males produce better quality ejaculates, and that some males are consistently better in competition (Dzuik, 1996), but Wade and Chang (1995) showed that best quality males may not always be most successful during sperm competition.

Haig and Bergstrom (1995) have suggested a similar kind of potential genetic benefit associated with polyandry. They point out that if some haploid control occurs, segregation distortion (SD) could also occur, and there is ample evidence for both (e.g. Erickson, 1990; Crow, 1979). If SD reduces ejaculate competitiveness, as it would when competition proceeds under the raffle principle, multiply mating females would effectively select against SD, with competitive benefits to their sons. In addition, if SD reduced overall fitness, selection would also favour polyandrous females. Benefits of this nature are likely to be transitory however because of strong selection for suppressor alleles (e.g. Wilkinson et al., 1998).

iii) Cryptic Female Choice & Genetic Incompatibility

The idea that multiple mating may function as a means of promoting cryptic female choice has the potential to explain the evolution of polyandry across a wide range of taxa (Eberhard, 1996). For example, male genitalia and/or copulatory behaviour may be used to continue courtship by stimulating the female during copulation, following which females bias paternity toward favoured males (Eberhard, 1996). Again, this hypothesis assumes that females gain fitness benefits via their sons. Nevertheless, while cryptic female choice has been documented in a number of taxa (e.g. Sakaluk, 1997; Reyer et al., 1999; Ward, 2000), in practice it is often difficult to eliminate alternative explanations for differential male fertilisation success.

Females may also mate multiply to promote the success of post-copulatory mechanisms for focusing reproductive investment in genetically viable offspring, where offspring viability is dependent on parental genetic compatibility. Female reproductive effort is wasted if ova are fertilised by genetically incompatible males that result in inviable zygotes. Zeh and Zeh (1996, 1997) point out that genetic incompatibility may be a widespread consequence of intragenomic conflict, for example resulting from cellular endosymbionts, transposable elements, segregation distorters, maternal effect lethals and genomically imprinted genes (see also Tregenza and Wedell, 2000). Incompatibility of maternal and paternal genomes may therefore select for polyandry if females are able to select sperm of genetically compatible males and/or reallocate investment to the most viable offspring produced. There is growing evidence that females may be capable of sperm selection based on genetic similarity or compatibility of males (e.g. Bishop et al., 1996; Olsson et al., 1996; Wilson et al., 1997; Stockley, 1999), although not in all taxa (e.g. Stockley, 1997a; Birkhead et al., 1999). Results of recent studies of sperm selection in relation to major histocompatibility (MHC) loci in mice suggest that selection for compatibility occurs at the ova, based on MHC haplotypes, and ensures heterozygote excess (Wedekind et al., 1996; Rulicke et al., 1998). If broadly applicable, these results may be significant also because they suggest a more general potential benefit for multiple mating than reducing costs of inbreeding—that is, promotion of offspring genetic quality with respect to disease resistance (review in Penn and Potts, 1999).

Genetic incompatibility appears the most likely explanation for the evolution of polyandry in pseudoscorpions. Here, polyandrous females have lower embryo mortality rates than monandrous females, suggesting male/female genetic incompatibility is a major cause of embryo mortality in this species (Zeh, 1997). Female pseudoscorpions also discriminate against previous mates, refusing to accept their sperm packages (Zeh et al., 1998). Tregenza and Wedell (1998) also found evidence to suggest that genetic incompatibility effects select for female multiple mating in a cricket, *Gryllus bimaculatus*. In this study, hatching success of eggs increased with the number of different males mated, while overall number of copulations and variation in male genetic quality were controlled for. Similar general effects have been proposed to occur in vertebrate taxa, possibly in relation to costs of inbreeding. For example, Kempenaers et al. (1999) recently reported that mixed paternity egg clutches of tree swallows (*Tachycineta bicolor*) have higher hatching success than clutches without extra-pair young. Since all unhatched eggs were fertilized, this result suggests that genetic incompatibility may be an important factor determining embryo mortality in this population, and hence that polyandry may be functioning to reduce offspring mortality.

iv) Selection Arenas

Multiple mating may also be beneficial for females where it increases the chances of producing genetically variable offspring for selection via post-fertilisation mechanisms, or selection arenas (Stearns, 1987). Such mechanisms are potentially costly, particularly with prolonged parental investment, as occurs in birds and mammals. Forbes and Mock (1998) present theoretical evidence that post-partum screening for offspring quality is most likely to be affordable under circumstances where there are large cohorts of evenly matched offspring with exaggerated variation in quality, and cheap sibling rivalry. In certain marsupial and primitive eutherian mammals, the production of large multiply sired litters of extremely altricial young appears consistent with these predictions. For example, multiple mating in common shrews followed by offspring competition apparently functions to increase mean genetic quality of surviving young (Stockley et al., 1993; Stockley and Macdonald, 1998).

v) Increased Offspring Diversity

Increasing the genetic diversity of offspring via multiple mating may be another mechanism for avoiding offspring reduction. Although it was initially suggested that promiscuity could bring little additional genetic variation to a clutch/litter relative to the large diversity of a single ejaculate (Williams, 1975), this need not be true (e.g. Yasui, 1998). Nevertheless, the conditions under which increased offspring diversity in and of itself is the selective advantage that drives female multiple mating seem limited (Yasui, 1998; Hosken and Blackenhorn, 1999). Reduced competition between half-sibs relative to full-sibs or increased co-operation between less related individuals may lead to a diversity advantage (Yasui, 1998), as may ecological conditions that lead to a concave up fitness function and hence select for increased variance, such as an expanding population (Hosken and Blanckenhorn, 1999). Potential examples of the former are seen among Hymenoptera (e.g. Crozier and Consul, 1976), and in honeybees, Fuchs and Moritz (1998) suggested that if specialist alleles are rare, then colony level selection could favour multiple mating to ensure that necessary but rare specialist genotypes are present. While this may explain the large numbers of copulations recorded in some social insects (>30 partners in some bees; see Fuchs and Moritz, 1998 and references therein), it seems to have less general applicability. The other benefit is parasite mediated, as colonies with greater genetic diversity (e.g. founded by polyandrous queens) are less susceptible to infection (Baer and Schmid-Hempel, 1999). This is essentially analogous to the MHC discussion above. However, in 'nursery' situations, increased diversity in a litter/brood may lead to increased rather than reduced competition (Birkhead and Møller, 1992; Mock and Parker, 1997; see above).

IV. OTHER FACTORS

It has also been suggested as alternatives to possible polyandry benefits that female multiple mating may also be merely a correlated response to selection on male mating frequency (Halliday and Arnold, 1987). Halliday and Arnold (1987) first suggested female multiple mating may be due to a genetic correlation between males and females arising as a consequence of pleiotropic gene action, that is, females mate repeatedly because genetic coupling has not allowed them to diverge from male promiscuity. However, recent work on *Drosophila* provides no evidence for genetic coupling (Sgró et al., 1998).

In other cases, polyandry itself may function as a form of mate assessment. For example, Colwell and Oring (1989) found that female spotted sandpipers copulated with extra-pair males when egg laying was completed, and in the subsequent breeding season paired with a male that they had performed an EPC with. Similarly, Wagner (1991) argued that since female razorbills (*Alca torda*) engage in EPC's during periods of infertility, they used the EPC's as a means of mate appraisal. There is also some evidence for mate appraisal driven polyandry in penguins (Schwartz et al., 1999).

Females will presumably obtain future benefits of some nature (e.g. decreased mate finding costs) from this mate evaluation.

DISCUSSION

Clearly, there exist many diverse ways in which females can benefit from polyandry. Moreover, benefits (and costs) may be manifest in more than one life-history component. Importantly, females of many taxa benefit from multiple mating via means other than simply increasing the number of offspring produced in a given reproductive attempt. This is unsurprising because female ability to influence the quality of their offspring is associated with an enormous diversity and complexity of life history strategies (Stearns, 1992; Roff, 1992), and hence different ways to benefit from polyandry. By acknowledging fundamental differences between male and female reproductive strategies—in particular that female reproductive strategies involve complex trade-offs with offspring quality and investment in future reproductive attempts, such that the number of offspring produced in any given reproductive attempt does not always relate simply to lifetime reproductive success—we can begin to make sense of the many and diverse potential benefits of polyandry. The resulting shift in emphasis is from explaining why females should remate at all, to explaining broader interspecific patterns of variation in remating behaviour and the costs and benefits of polyandry.

As yet, it is only possible to identify the broadest associations between benefits of polyandry and life history variation. Examples of increased female longevity associated with multiple mating have so far only been reported in certain insects, females of which presumably experience high adult mortality risk. Our review also suggests that females of species that invest in producing very large numbers of offspring are probably more susceptible to sperm limitation than those which produce relatively few. Examples of multiple mating in association with sperm limitation are found most frequently among fishes and insects (e.g. Nakamura and Kramer, 1982; Ridley, 1988). Fecundity related benefits of increased offspring production may also be more significant for taxa in which females produce large numbers of offspring with relatively low investment in each, as evidenced by a recent review of benefits of polyandry in insects (Arnqvist and Nilsson, 2000). Here, we argue that a useful distinction could be made between those cases in which female fecundity is increased as a result of polyandry (sensu Bateman, 1948), and those in which females benefit by avoiding reductions to the number of offspring in which they have already invested. The latter is apparently more widespread, including examples of females ensuring that all their ova are fertilised, minimising offspring mortality due to parental genetic incompatibility, and avoiding infanticide. In contrast, examples where females instead promote offspring mortality after an initial overproduction have been reported less frequently (section IIIiv), but nonetheless fit theoretical expectations that females adopting this strategy should produce relatively cheap or altricial young of variable quality.

Life history strategies involving high early reproductive effort or short reproductive seasons may also be more commonly associated with benefits of polyandry via the provision of nuptial gifts, which enable females to increase their rate of reproduction by improving foraging efficiency. Such a strategy may be favoured particularly under conditions of high extrinsic adult mortality, when it is important for females to reproduce quickly. Similarly, utilisation of nuptial gifts for the female's own somatic maintenance should be favoured more when mortality of adults due to random or extrinsic causes is low relative to that of offspring, or when environmental nutrients are scarce (e.g. Simmons and Gwynne, 1993), whereas high extrinsic mortality caused by predation may instead favour the evolution of nuptial gifts that can be utilised to increase survival (e.g. Gonzáles et al., 1999). Unfortunately, studying ejaculate nutrient donations is in many ways problematic (discussed in Pitnick et al., 1997). For example, techniques using radio-labels may detect donated nutrients at levels that are biologically insignificant, or conversely donations may be qualitative rather than quantitative.

In line with general life history trends, our review suggests that females of species that invest in producing relatively few high quality offspring may be generally more likely to remate for benefits associated with improving offspring quality rather than increasing their numbers. Much evidence for remating decisions based on male quality, for example, comes from avian studies (sections III i & IV). Selection to minimise mortality risk to offspring is also likely to be strong if females invest heavily in offspring quality, which may explain why infanticide avoidance is apparently widespread in certain mammal groups, although not why it isn't more widespread in other taxa.

The extent to which offspring quality depends on the amount of parental investment received, and/or on genetic quality, may also be an important consideration. If offspring competitive ability is strongly dependent on the amount of parental investment received, then females may be more likely to remate in order to minimise energetic or time costs and hence maximise investment in their offspring. For example, the amount of maternal investment received is often an important predictor of offspring reproductive success in large polygynous mammals, which typically give birth to single offspring (Clutton-Brock, 1988; Carranza, 1996), hence we might expect multiple mating by these females to be associated with cost minimisation or gaining access to resources. Where offspring genetic quality is variable, females may instead or additionally be selected to remate for genetic benefits. Offspring competitive ability is apparently variable due to costs of inbreeding in certain populations of reptiles and small mammals in which female promiscuity is common (Olsson et al., 1996; Stockley et al., 1993), and benefits associated with genetic compatibility may be more important generally in small or inbred populations with low dispersal (e.g. Bishop et al., 1996; Olsson et al., 1996). Furthermore, it is likely that large investment in each offspring will favour multiple mating as a means of avoiding genetic incompatibilities (Zeh, 1997), and it seems likely that more instances will be found as researchers become increasingly aware of this possibility.

Other hypotheses to explain benefits of polyandry, such as those suggesting benefits associated with heritable male fertilisation success (e.g. section III ii), might apply across diverse taxa with very different life history strategies. As yet, insufficient data are available to test the generality of such explanations, or indeed whether other potential benefits might be more taxonomically widespread than suggested by the current review. Future studies might address these questions both by expanding investigation of broadly applicable hypotheses to more diverse taxa, and by formulating more explicit predictions about expected costs and benefits of polyandry according to the life history strategies of key study species.

Our review also highlights the different ways in which polyandry is often associated with sexual conflict (Parker, 1979). Conflict over mating is most likely to arise when the costs of remating for females are high relative to the benefits, such that females are reluctant to remate. Such circumstances may favour male persistence or coercion, with the result that females mate multiply in order to minimise costs (e.g. Clutton-Brock and Parker, 1995), or the provision of male nuptial gifts to encourage female remating (Gwynne, 1997). In both cases, females do not benefit from initiating polyandry *per se*, but may remate as a consequence of male adaptations favoured under sexual conflict. Similarly, exaggerated male courtship displays and showy traits may also be viewed as arising from selection under sexual conflict, with the function of persuading females to mate when it is not always in their best interests to do so (Holland and Rice, 1998). Conversely, when females benefit from initiating polyandry in the absence of male adaptations designed to coerce them into doing so, sexual conflict over sperm competition often results in male adaptations to limit the expression of female remating behaviour (Parker, 1979; Stockley, 1997b). Hence, paradoxically, it appears that polyandry may often be favoured under sexual conflict when it is most costly to females, and constrained as a result of similar selection pressures when it is of most benefit. A further important consideration here from a life history perspective is the extent to which apparent benefits for females may ultimately prove costly. For example, an increase in fecundity associated with multiple mating could result from an increase in nutrients received, and/or from substances contained in the ejaculate or nuptial gift that increase female fecundity. In the former case females may benefit from an increase in offspring production, but in the latter they are likely to suffer long term costs if prevailing male interests cause them to exceed an optimum level of reproductive effort. A potentially important distinction exists here with respect to the timing of male parental investment. Sustained male investment in the offspring may permit an increase in optimal clutch size for females, whereas the provision of nuptial gifts at the time of copulating in general may be less likely to result in mutual benefits. Because there is potential for sexual conflict and associated increased costs of reproduction when females remate, long term measures of reproductive success are critical to evaluate potential costs and benefits of polyandry, particularly with respect to offspring numbers, because evidence of conflict may often only be manifest over a female's lifetime.

In conclusion, we believe the life history perspective promoted here will ultimately lead to clearer interpretation of current evidence regarding in the benefits of polyandry among different taxa (e.g. Chao, 1997). In contrast with the Bateman paradigm, this approach shifts emphasis from considering why females mate multiply, to focus instead on explaining variation in benefits at a broader level. Importantly, this approach also highlights the importance of sexual conflict as a selective force mediating the evolution and expression of female multiple mating behaviour. Although it is only possible to begin making the broadest generalisations as yet, we believe that more detailed consideration of the likely consequences of variation in life history strategies will offer much potential in the future to further understanding of female mating behaviour.

SUMMARY

While there is accumulating evidence for diverse potential benefits of polyandry, the extent to which general hypotheses might apply across taxa is still unclear. Here, we begin to address this question by reviewing current evidence in relation to different life history components. This approach shifts emphasis away from the Bateman paradigm and explaining why polyandry should occur at all, to begin looking for general trends in the kinds of benefits associated with polyandry across diverse taxa. Benefits are considered under broad categories including effects on female longevity and the number and quality of their surviving offspring. Several general trends emerge from this treatment. Benefits of polyandry that increase female lifespan and/or fecundity have been most commonly reported in invertebrate taxa with presumed high rates of mortality and high reproductive output. Fertility benefits are also most commonly found in taxa where females invest in producing large numbers of offspring, whereas benefits associated with improving offspring quality and survival are more commonly reported where females invest in relatively few offspring per reproductive attempt. It is noted that several benefits often associated with maximising offspring numbers, such as fertility benefits, and avoiding genetic incompatibility or infanticide, actually function to avoid reductions to the clutch size in which females have already invested rather than to increase numbers *per se*. Other potential benefits of polyandry, such as those associated with heritable male fertilisation success, might apply across taxa with diverse life history strategies, although there is currently insufficient evidence to determine how widespread these benefits might be. Our review also incorporates new evidence linking polyandry with sexual conflict. Because sexual conflict may increase costs of reproduction for females, we conclude that it is important to consider both costs and benefits of polyandry in the long term wherever possible.

Acknowledgments

We wish to thank Trent Garner for discussion of paternity detection and genetic diversity, Oliver Martin, Christoph Vorburger, Paul Ward, Scott Pitnick, Brett Holland, Tom Tregenza and Nina Wedell for insightful comments on earlier versions of this manuscript. DJH was supported by the Swiss National Science Foundation and the Alexander Von Humboldt Stiftung.

REFERENCES

Alcock, J., 1989, *Animal Behavior: An Evolutionary Perspective*, 4th edition, Sinaur & Associates, Sunderland, Mass.

Alcock, J., Barrows, E. M., Gordh, G., Hubbard, L. J., Kirkendall, L., Pyle, D. W., Ponder, T. L., and Zalom, F. G., 1978, The ecology and evolution of male reproductive behaviour in the bees and wasps, *Zool. J. Linn. Soc.* **64:**293–326.

Arnqvist, G., 1989, Multiple mating in the water strider: mutual benefit or intersexual conflict?, *Anim. Behav.* **38:**749–756.

Arnqvist, G., 1997, The evolution of water strider mating systems: causes and consequences of sexual conflict, in: *The Evolution of Mating Systems in Insects and Arachnids* (J. C. Choe and B. J. Crespi, eds.), pp. 146–163, Cambridge University Press, Cambridge.

Arnqvist, G., and Rowe, L., 1995, Sexual conflict and arms races between the sexes: a morphological adaptation for control of mating in a female insect, *Proc. R. Soc. Lond., B* **261:**123–127.

Arnqvist, G., and Nilsson, T., 2000, The evolution of polyandry: multiple mating and female fitness in insects, *Anim. Behav.* **60:**145–164.

Baer, B., and Schmid-Hempel, P., 1999, Experimental variation in polyandry affects parasite loads and fitness in a bumble bee, *Nature* **397:**151–154.

Ball, M. A., and Parker, G. A., 1996, Sperm competition games: external fertilisation and 'adaptive' infertility, *J. Theor. Biol.* **180:**141–150.

Bateman, A. J., 1948, Intrasexual selection in *Drosophila*, *Heredity* **2:**349–368.

Birkhead, T. R., and Møller, A. P., 1992, *Sperm Competition in Birds: Evolutionary Causes and Consequences*, Academic Press, London.

Birkhead, T. R., and Møller, A. P., 1998, *Sperm Competition and Sexual Selection*, Academic Press, London.

Birkhead, T. R., Møller, A. P., and Sutherland, W. J., 1993, Why do females make it so difficult for males to fertilize their eggs?, *J. Theor. Biol.* **161:**51–60.

Birkhead, T. R., Martinez, J. G., Burke, T., and Froman, D. P., 1999, Sperm mobility determines the outcome of sperm competition in the domestic fowl, *Proc. R. Soc. Lond., B* **266:** 1759–1764.

Bishop, J. D. D., Jones, C. S., and Noble, L. R., 1996, Female control of paternity in the internally fertilizing compound ascidian *Diplosoma listeriannum*. II. Investigation of male mating success using RAPD markers, *Proc. R. Soc. Lond., B* **263:**401–407.

Boomsma, J. J., 1996, Split sex ratios and queen-male conflict over sperm allocation, *Proc. R. Soc. Lond., B* **263:**697–704.

Borries, C., Launhardt, K., Epplen, C., Epplen, J. T., and Winkler, P., 1999, DNA analyses support the hypothesis that infanticide is adaptive in langur monkeys, *Proc. R. Soc. Lond., B* **266:**901–904.

Burpee, D. M., and Sakaluk, S. K., 1993, Repeated matings offset the costs of reproduction in female crickets, *Evol. Ecol.* **7:**240–250.

Briskie, J. V., Montgomerie, R., Põldmaa, T., and Boag, P. T., 1998, Paternity and paternal care in the polygynandrous Smith's longspur, *Behav. Ecol. Sociobiol.* **43:**181–190.

Brown, J. L., 1997, A theory of mate choice based on heterozygosity, *Behav. Ecol.* **8:**60–65.

Brown, W. D., 1997, Courtship feeding in tree crickets increases insemination and female reproductive lifespan, *Anim. Behav.* **54:**1369–1382.

Carranza, J., 1996, Sexual selection for male body mass and the evolution of litter size in mammals, *Am. Nat.* **148:**81–100.

Chao, L., 1997, Evolution of polyandry in a communal breeding system, *Behav. Ecol.* **8:**668–674.

Cicirello, D. M., and Wolff, J. O., 1990, The effects of mating on infanticide and pup discrimination in white-footed mice, *Behav. Ecol. Sociobiol.* **26:**275–279.

Clutton-Brock, T. H., 1988, *Reproductive Success*, The University of Chicago Press, Chicago.

Clutton-Brock, T. H., and Parker, G. A., 1995, Sexual coercion in animal societies, *Anim. Behav.* **49:**1345–1365.

Colwell, M. A., and Oring, L. W., 1989, Extra-pair maing in the spotted sandpiper: a female mate aquisition tactic, *Anim. Behav*. **38:**675–684.

Crow, J. F., 1979, Genes that violate Mendel's laws, *Sci. Am*. **240:**134–146.

Crozier, R. H., and Consul, P. C., 1976, Conditions for genetic polymorphism in social Hymenoptera under selection at the colony level, *Theor. Pop. Biol*. **10:**1–9.

Davies, N. B., 1985, Cooperation and conflict among dunnocks, *Prunella modularis*, in a variable mating system, *Anim. Behav*. **33:**628–648.

Davies, N. B., Hartley, I. R., Hatchwell, B. J., and Langmore, N. E., 1996, Female control of copulations to maximize male help: a comparison of polygynandrous alpine accentors, *Prunella collaris*, and dunnocks, *P. modularis*, *Anim. Behav*. **51:**27–47.

Dzuik, P. J., 1996, Factors that influence the proportion of offspring sired by a male following heterospermic insemination, *Anim. Reprod. Sci*. **43:**65–88.

Eberhard, W. G., 1996, *Female Control: Sexual Selection by Cryptic Female Choice*, Princeton University Press, Princeton.

Eisner, T., Eisner, M., Rossini, C., Iyengar, V. K., Roach, B. L., Benedikt, E., and Meinwald, J., 2000, Chemical defense against predation in an insect egg, *Proc. Natl. Acad. Sci. USA* **37:**1634–1639.

Emlen, S. T., and Oring, L. W., 1977, Ecology, sexual selection and the evolution of mating systems, *Science* **197:**215–223.

Erickson, R. P., 1990, Post-meiotic gene expression, *Trends Genet*. **8:**264–269.

Fairbain, D. J., 1993, Costs of loading associated with mate-carrying in the waterstrider, *Aquarius remigis*, *Behav. Ecol*. **4:**224–231.

Festa-Bianchet, M., Gallard, J. M., and Jorgenson, J. T., 1998, Mass and density dependent reproductive success and reproductive costs in a capital breeder, *Am. Nat*. **152:**367–379.

Forbes, S. L., and Mock, D. W., 1998, Parental optimism and progeny choice: when is screening for offspring quality affordable?, *J. Theor. Biol*. **192:**3–14.

Fjerdingstad, E. J., and Boomsma, J. J., 1998, Multiple mating increases the sperm stores of *Atta colombica* leafcutter ant queens. *Behav. Ecol. Sociobiol*. **9:**257–261.

Fox, C. W., 1993, Multiple mating, lifetime fecundity and female mortality of the bruchid beetle, *Callosobruchus maculatus* (Coleoptera: Bruchidae), *Funct. Ecol*. **7:**203–208.

Fuchs, S., and Moritz, R. F. A., 1998, Evolution of extreme polyandry in the honeybee *Apis mellifera* L., *Behav. Ecol. Sociobiol*. **9:**269–275.

Gibson, R. M., and Jewell, P. A., 1982, Semen quality, female choice, and multiple mating in domestic sheep: a test of Triver's sexual competence hypothesis, *Behaviour* **80:**9–31.

Gonzáles, A., Rossini, C., Eisner, M., and Eisner, T., 1999, Sexually transmitted chemical defense in a moth (*Utetheisa ornatrix*), *Proc. Natl. Acad. Sci. USA* **96:**5570–5574.

Gwynne, D., 1984, Courtship feeding increases female reproductive success in bushcrickets, *Nature* **307:**361–363.

Gwynne, D., 1997, The evolution of edible 'sperm sacs' and other forms of courtship feeding in crickets, katydids and their kin (Orthoptera: Ensifera) in: *The Evolution of Mating Systems in Insects and Arachnids* (J. C. Choe and B. J. Crespi, eds.), pp. 146–163, Cambridge University Press, Cambridge.

Haig, D., and Bergstrom, C. T., 1995, Multiple mating, sperm competition and meiotic drive, *J. Evol. Biol*. **8:**265–282.

Halliday, T., and Arnold, S. J., 1987, Multiple mating by females: a perspective from quantitative genetics. *Anim, Behav*. **35:**939–941.

Holland, B., and Rice, W. R., 1998, Perspective: Chase-away sexual selection: antagonistic seduction versus resistance, *Evolution* **52:**1–7.

Hosken, D. J., and Blanckenhorn, W. U., 1999, Female multiple mating, inbreeding avoidance and fitness: it's not only the magnitude of costs and benefits that count, *Behav. Ecol.* **10:**462–464.
Hosken, D. J., Blanckenhorn, W. U., and Ward, P. I., 2000, Developmental stability in yellow dung flies (*Scathophaga stercoraria*): fluctuating asymmetry, heterozygosity and environmental stress, *J. Evol. Biol.* **13:**919–926.
Hosken, D. J., Garner, T. W. J., and Ward, P. I., 2001, Sexual conflict selects for male and female characters, *Current Biology* **11:**489–493.
Hrdy, E., 1977, *The Langurs of Abu*, Harvard University Press, Cambridge, Mass.
Jennions, M. D., and Petrie, M., 2000, Why do females mate multiply? A review of the genetic benefits, *Biol. Rev.* **75:**21–64.
Karlsson, B., 1998, Nuptial gifts, resource budgets, and reproductive output in a polyandrous butterfly, *Ecology* **79:**2931–2940.
Keller, L., and Reeve, H. K., 1995, Why do females mate with multiple males? The sexually selected sperm hypothesis. *Adv. Study Behav*. **24:**291–315.
Kempenaers, B., Congdon, B., Boag, P., and Robertson, R. J., 1999, Extrapair paternity and egg hatchability in tree swallows: evidence for the genetic compatability hypothesis?, *Behav. Ecol.* **10:**304–311.
LaMunyon, C., 1997, Increased fecundity, as a function of multiple mating, in an arctiid moth, *Utetheisa ornatatrix, Ecol. Entomol.* **22:**69–73.
Lauer, M. J., 1996, Effect of sperm depletion and starvation on female mating behavior in the water strider *Aquarius remigis. Behav. Ecol. Sociobiol.* **38:**89–96.
Madsen, T., and Shine, R., 1998, Quantity or quality? Determinants of maternal reproductive success in tropical pythons (*Liasis fuscus*), *Proc. R. Soc. Lond., B* **265:**1521–1525.
Madsen, T., Shine, R., Loman, J., and Hakansson, T., 1992, Why do female adders copulate so frequently?, *Nature* **355:**440–441.
Markow, T. A., Gallagher, P. D., and Krebs, R. A., 1990, Ejaculate derived nutritional contribution and female reproductive success in *Drosophila mojavensis*, *Funct. Ecol.* **4:**67–73.
Mock, D. W., and Parker, G. A., 1997, *The Evolution of Sibling Rivalry,* Oxford University Press, Oxford.
Møller, A. P., 1988, Infanticidal and anti-infanticidal strategies in the barn swallow *Hirundo rustica*, *Behav. Ecol. Sociobiol.* **22:**365–371.
Nakatsuru, K., and Kramer, D. L., 1982, Is sperm cheap? Limited fertility and female choice in the lemon tetra (Pisces, Characidae), *Science* **216:**753–755.
Nakamura, M., 1998, Multiple mating and cooperative breeding in polygyandrous alpine accentor. I. Competition among females, *Anim. Behav.* **55:**259–275.
Olsson, M., Shine, R., Madsen, T., Gullberg, A., and Tegelstrøm, H., 1996, Sperm selection by females, *Nature* **383:**585.
Partridge, L., 1983, Non-random mating and offspring fitness, in: *Mate Choice* (P. Bateson, ed.), pp. 227–255, Cambridge University Press, Cambridge.
Parker, G. A., 1979, Sexual selection and sexual conflict, in: *Sexual Selection and Reproductive Competition in Insects* (M. S. Blum and N. A. Blum, eds.), pp. 123–166, Academic Press, London.
Penn, D. J., and Potts, W. K., 1999, The evolution of mating preferences and major histocompatibility complex genes, *Am. Nat.* **153:**145–164.
Petrie, M., Doums, C., and Møller, A. P., 1998, The degree of extra-pair paternity increases with genetic diversity, *Proc. Natl. Acad. Sci. USA* **95:**9390–9395.
Pitnick, S., and Markow, T. A., 1994, Male gametic strategies: sperm size, testes size, and the allocation of ejaculate among successive mates by the sperm-limited fly *Drosophila pachea* and its relatives, *Am. Nat.* **143:**785–819.

Pitnick, S., Markow, T. A., and Reidy, M. F., 1991, Transfer of ejaculates and incorporation of male derived substances by females of the Nannopteran species group (Diptera: Drosophila), *Evolution* **45:**774–780.

Pitnick, S., Spicer, G. S., and Markow, T., 1997, Phylogenetic examination of female incorporation of ejaculate in *Drosophila. Evolution* **51:**833–845.

Radwan, J., 1998, Heritability of sperm competition success in the bulb mite, *Rhizoglyphus robini, J. Evol. Biol.* **11:**321–327.

Reyer, H.-U., Frei, G., and Som, C., 1999, Cryptic female choice: frogs reduce clutch size when amplexed by undesired males, *Proc. R. Soc. Lond., B* **266:**2101–2107.

Ridley, M., 1988, Mating frequency and fecundity in insects, *Biol. Rev.* **63:**509–549.

Roff, D. A. 1992, *The Evolution of Life Histories: Theory and Analysis*, Chapman & Hall, London.

Rulicke, T., Chapuisat, M., Homberger, F. R., Macas, E., and Wedekind, C., 1998, MHC-genotype of progeny influenced by parental infection, *Proc. R. Soc. Lond., B* **265:**711–716.

Sax, A., Hoi, H., and Birkhead, T. R., 1998, Copulation rate and sperm use by female bearded tits, *Panurus biarmicus, Anim. Behav.* **56:**1199–1204.

Sakaluk, S. K., 1997, Cryptic female choice predicated on wing dimorphism in decorated crickets, *Behav. Ecol.* **8:**326–331.

Savalli, U. M., and Fox, C. W., 1999, The effects of male mating history on paternal investment, fecundity and female remating in the seed beetle *Callosobruchus maculatus*, *Funct. Ecol.* **13:**169–177.

Schwartz, M. K., Boness, D. J., Schaeff, C. M., Majluf, P., Perry, E. A., and Fleischer, R. C., 1999, Female-solicited extrapair matings in Humboldt penguins fail to produce extrapair fertilizations, *Behav. Ecol.* **10:**242–250.

Sgró, C. M., Chapman, T., and Partridge, L., 1998, Sex specific selection on time to remate in *Drosophila melanogaster*, *Anim. Behav.* **56:**1267–1278.

Simmons, L. W., 1988, The contribution of multiple mating and spermatophore consumption to the lifetime reproductive success of female field crickets, *Ecol. Entomol.* **13:**57–69.

Simmons, L. W., and Bailey, W. J., 1990, Resource influenced sex roles of zaprochiline tettigoniids (Orthoptera: Tettigonidae), *Evolution* **44:**1853–1868.

Simmons, L. W., and Gwynne, D., 1993, Reproductive investment in bushcrickets: the allocation of male and female nutrients to offspring, *Proc. R. Soc. Lond., B* **252:**1–5.

Stearns, S. C., 1987, The selection arena hypothesis, in: *The Evolution of Sex and its Consequences* (S. C. Stearns, ed.), pp. 337–379, Birkhäuser Verlag, Basel.

Stearns, S. C., 1992, *The Evolution of Life Histories*, Oxford University Press, Oxford.

Stockley, P., 1997a, No evidence for sperm selection by female common shrews, *Proc. R. Soc. Lond., B* **264:**1497–1500.

Stockley, P., 1997b, Sexual conflict resulting from adaptations to sperm competition. *Trends Ecol. Evol.* **12:**154–159.

Stockley, P., 1999. Sperm selection and genetic incompatibility: does relatedness of mates affect male success in sperm competition?, *Proc. R. Soc. Lond., B* **266:**1663–1669.

Stockley, P., and Macdonald, D. W., 1998, Why do female common shrews produce so many offspring?, *Oikos* **83:**560–566.

Stockley, P., Searle, J. B., Macdonald, D. W., and Jones, C. S., 1993, Female multiple mating behaviour in the common shrew as a strategy to reduce inbreeding., *Proc. R. Soc. Lond., B* **254:**173–179.

Thornhill, R., 1976, Sexual selection and nuptial feeding behavior in *Bittacus apicalis* (Insecta: Mecoptera). *Am. Nat.* **110:**529–548.

Thornhill, R., 1978, Sexually selected predatory and mating behavior of the hanging fly, *Bittacus stigmaterus* (Mecoptera: Bittacidae). *Annals Entomol. Soc. Am.* **71:**597–601.

Tregenza, T., and Wedell, N., 1998, Benefits of multiple mating in the cricket *Gryllus bimaculatus*, *Evolution* **52:**1726–1730.

Tregenza, T., and Wedell, N., 2000, Genetic compatability, mate choice and patterns of parentage, *Mol. Ecol.* **9:**1013–1027.

Viera, C., Pasyukova, E. G., Zeng, Z.-B., Hackett, J. B., Lyman, R. F., and Mackay, T. F. C., 2000, Genotype-environment interactions for quantitative trait loci affecting life span in *Drosophila melanogaster*, *Genetics* **154:**213–227.

Vøllestad, L. A., Hindar, K., and Møller, A. P., 1999, A meta-analysis of fluctuating asymmetry in relation to heterozygosity, *Heredity* **83:**206–218.

Wade, M. J., and Chang, N. W., 1995, Increased male fertility in *Tribolium confusum* beetles after infection with the intra-cellular parasite *Wolbachia*, *Nature* **373:**72–74.

Wagner, R. H., 1991, The use of extrapair copulation for mate appraisal by razor bills, *Alca torda*, *Behav. Ecol.* **2:**198–203.

Ward, P. I., 2000, Cryptic female choice in the yellow dung fly *Scathophaga stercoraria* (L.), *Evolution* **54:**1680–1686.

Warner, R. R., Shapiro, D. Y., Marcanato, A., and Petersen, C. W., 1995, Sexual conflict: males with highest mating success convey the lowest fertilisation benefits to females, *Proc. R. Soc. Lond., B* **262:**135–139.

Weatherhead, P. J., Dufour, K. W., Lougheed, S. C., and Eckert, C. G., 1999, A test of the good-genes-as-heterozygosity hypothesis using red-winged blackbirds, *Behav. Ecol.* **10:**619–625.

Wedell, N., 1996, Mate quality affects reproductive effort in a paternally investing species, *Am. Nat.* **148:**1075–1088.

Wedekind, C., Chapsuit, M., Macas, E., and Rulicke, T., 1996, Non-random fertilization in mice correlates with MHC and something else, *Heredity* **77:**400–409.

Wildt, D. E., Bush, M., Goodrowe, K. L., Packer, C., Pusey, A. E., Brown, J. L., Joslin, P., and O'Brien, S. J., 1987, Reproductive and genetic consequences of founding isolated lion populations, *Nature* **329:**328–331.

Wiley, R. H., and Poston, J., 1996, Indirect mate choice, competition for mates and the coevolution of the sexes, *Evolution* **50:**1371–1381.

Williams, G. C., 1975, *Sex and Evolution,* Princeton University Press, Princeton.

Wilkinson, G. S., Presgraves, D. C., and Crymes, L., 1998, Male eye span in stalk-eyed flies indicates quality by meotic drive suppression, *Nature* **391:**276–279.

Wilson, N., Tubman, S., Eady, P., and Robertson, G. W., 1997, Female genotype affects male success in sperm competition, *Proc. R. Soc. Lond., B* **264:**1491–1495.

Yasui, Y., 1998, The 'genetic benefits' of female multiple mating reconsidered, *Trends Ecol. Evol.* **13:**246–250.

Zeh, J. A., 1997, Polyandry and enhanced reproductive success in the harlequin-beetle-riding psuedoscorpion, *Behav. Ecol. Sociobiol.* **40:**111–118.

Zeh, J. A., Newcomer, S. D., and Zeh, D. W., 1998, Polyandrous females discriminate against previous mates, *Proc. Natl. Acad. Sci. USA* **95:**13732–13736.

Zeh, J. A., and Zeh, D. W., 1996, The evolution of polyandry. I. Intragenomic conflict and genetic incompatibility. *Proc. R. Soc. Lond., B* **263:**1711–1717.

Zeh, J. A., and Zeh, D. W., 1997, The evolution of polyandry. II. Post-copulatory defence against genetic incompatibility. *Proc. R. Soc. Lond., B* **264:**69–75.

Index

MIX
Papier aus verantwortungsvollen Quellen
Paper from responsible sources
FSC® C105338

If you have any concerns about our products,
you can contact us on
ProductSafety@springernature.com

In case Publisher is established outside the EU,
the EU authorized representative is:
Springer Nature Customer Service Center GmbH
Europaplatz 3, 69115 Heidelberg, Germany

Printed by Libri Plureos GmbH
in Hamburg, Germany